P9-EDU-519

More Praise For

MAKE IT IN AMERICA

America used to make things. Americans were the innovators, the undisputed world economic leaders, and, with our success, we built a thriving, high-achieving middle class. Now, with policies that are indifferent or hostile to domestic manufacturing, and with other nations on the rise, America's long-term prosperity is at risk. Our only chance to turn things around is to remember, revive, and revolutionize what made us great for so long—the manufacturing sector.

"Liveris takes a hard look at the current state of manufacturing and the nature of the twenty-first century global economy. His dramatic call to action—to revive the American economy by reinventing the American manufacturing sector—is cogent and heartfelt. *Make It in America* provides a valuable strategic framework for action by governmental, business, and community leaders, and it deserves to be a part of the ongoing national discussion about the country's economic priorities and its future."

—James W. Owens
Chairman and CEO, Caterpiller Inc.

"Andrew Liveris is one of America's most active and thoughtful CEOs. I'm convinced that we must have more savings, investments, and exports to offset the twin deficits (fiscal and trade), which certainly contributed to the worst global recession we've experienced since the 1930s. A vibrant, globally competitive manufacturing sector must be part of our 'economic fix,' and Andrew offers numerous suggestions on getting our mojo back. This is a call to action!"

—Jim Quigley
CEO, Deloitte Touche Tohmatsu Limited

MAKE IT
IN
AMERICA

MAKE IT IN AMERICA

THE CASE FOR RE-INVENTING THE ECONOMY

Andrew Liveris

CEO, The Dow Chemical Company

John Wiley & Sons, Inc.

Published by John Wiley & Sons, Inc., Hoboken, New Jersey.
Published simultaneously in Canada.

Library of Congress Cataloging-in-Publication Data:

Liveris, Andrew.
Make it in America : the case for re-inventing the economy / Andrew Liveris.
p. cm.
Includes bibliographical references and index.
ISBN 978-0-470-93022-9 (cloth); ISBN 978-1-118-01938-2 (ebk);
ISBN 978-1-118-01939-9 (ebk); ISBN 978-1-118-01940-5 (ebk)
1. Industrial policy–United States. 2. Manufactures–United States. 3. Economic forecasting–United States. 4. United States–Economic conditions–2009– I. Title.
HD3616.U47L58 2010
330.973–dc22 2010045654

Printed in the United States of America

10 9 8 7 6 5 4 3 2 1

To the women and men of The Dow Chemical Company,
past and present, who have positively impacted
countless lives around the world with their actions,
their products, and their values.

Contents

CONTENTS

Preface

I'm Andrew Liveris. I've been an employee of The Dow Chemical Company for more than 30 years. For the past six, I've been its Chairman and CEO. I was born in Australia, in the small town of Darwin, in 1954.

Throughout my childhood, I always loved school. Chemistry, in particular. There was something fascinating to me about manipulating elements to produce vastly different compounds. The idea of working at the molecular level was at once exciting and mystifying. When I got to University of Queensland, I knew right away that I would study chemical engineering.

Upon graduation, I came to The Dow Chemical Company as a chemical engineer. I didn't join Dow because it was a chemical company. I joined it because it was an American company. I joined it because I believe deeply in the American model of enterprise, in its ability to improve lives around the world, in its capacity to be a beacon of energy and creativity.

And Dow is about as quintessentially American as you can get. It was founded in 1897 by an immigrant—an entrepreneur— just as the industrial revolution was getting underway.

Herbert Henry Dow, a Canadian chemical engineer, came to the small town of Midland, Michigan to collect samples from the Tittabawassee River. He found that he could extract bromine, an element used in products from flame retardants to gasoline, from the brine lakes in the area. After securing patents, he, along with 57 investors and $200,000, started The Dow Chemical Company.

Dow grew over the twentieth century to become one of the largest and most global corporations in the world. Today Dow employs more than 52,000 workers in 37 countries. We have 41 major research facilities, more than 5,000 products being manufactured at 214 sites, and sales in more than 160 countries. We are a Fortune 50 company.

I am deeply indebted to the country that allowed me to pursue my dreams. I worked for Dow in Australia, then for many years in Asia, and now, as CEO, I work out of Dow's headquarters, which have remained in Midland to this day.

As the head of one of the largest chemical companies in the world, I interact with manufacturers of countless kinds every day. They are my peers, my partners and, at times, my customers. Our conversations, I believe, give me a keen sense of the pressures and challenges they're facing, the changes they're making in response, and the opportunities they're seeing—and seizing. So I speak from personal and professional experience (and with no shortage of pride in my industry) when I say this:

The world is entering a golden age of manufacturing.

Now, that statement might surprise many in America. We've become painfully accustomed to the loss of jobs, the closing of plants, the shuttering of main streets in small towns and even some big cities. To many Americans, the word "manufacturing" feels associated with the past—with another era, a brighter and easier and more prosperous time.

So when I say manufacturing is about to enter a golden age, it might seem impossible to reconcile with the reality to which we've become accustomed. But if we lift our eyes from what's happening within our own borders and begin to look around the world, we can see that, globally, this is an incredibly exciting time in manufacturing—perhaps the most exciting in history.

Countries are investing extraordinary amounts of money and talent into expanding their capacity to create—and to build. In China, for example, cities with populations larger than New York's have sprouted up out of nowhere in order to meet the country's growing manufacturing sector. Of course, China is not alone in its mission to become the global leader in manufacturing. Developed and developing nations alike, from India and Germany to Brazil and Taiwan, are creating comprehensive national strategies to better compete in the global market. They are creating new industries that will be central to solving some of the world's most serious challenges. Indeed, countries all over the world see manufacturing as the key to their economic futures.

America's most successful global competitors are building substantial wealth by investing in highly advanced, highly specialized, high value-added manufacturing—building the semiconductors and microprocessors for our electronics; the wind turbines and solar cells for our energy needs; the advanced batteries and state-of-the-art medical devices that will remake our future. They are transforming what manufacturing means, and using its engines to transform their economies. It is truly stunning to watch this unfold.

There is no doubt that, for generations to come, economic success will be a direct product of the things we build. The question is, who will build them? Who will reap the rewards? Who will emerge as the economic leader of the twenty-first century?

In that race, the United States is falling rapidly behind.

Other nations have refocused and ramped up, but the United States has not. In the World Economic Forum's 2010–2011 Global Competitiveness rankings, the United States slid another two slots—from second to fourth. We have allowed our manufacturing base to deteriorate, and we haven't done nearly enough to revive it. To reinvent it.

We mourn the loss of manufacturing jobs in America. We recognize the pain it causes to workers and their families and their hometowns. We lament it. But we don't do much about it. Just as inaction led us to this point, inaction is keeping us here. We treat further losses as inevitable, even acceptable, and ignore their effect on our long-term success. We are talking *around* the problem, but very few people are actually talking *about* the problem.

This is where we are today. And we arrived at this moment without ever asking the most important question of all.

Does manufacturing matter to our future?

Yes, it does. It absolutely does.

If you picked up this book thinking it would be another conventional business book, I urge you to keep reading—I'm confident you'll conclude otherwise. If you picked it up thinking this was another long complaint by another CEO who wants nothing more than for government to back off, recede from the picture, do nothing, and let the markets rule, then I'm afraid you'll be disappointed.

That approach had its day. There was a time when companies could thrive, when entire enterprises could be built from the ground up on our shores, without much government intervention or assistance. We saw a vast prosperity arise, we saw innovation flourish, and many Americans concluded that government and business should keep their distance from one another. It's part of our national DNA.

If everything had remained static, it might have worked out just fine. But it didn't. And we ignore those changes at our peril.

This book is a call to action. It's based not just on my experience, but on the opinions of some of the leading experts in the field. Over the next several chapters I'll give you a better sense of where the United States is falling behind and why. I will talk about some of the lessons we can learn—the lessons I have learned—from our major competitors around the world. And, in the final chapters, I will lay out an agenda that, if adopted, could revive the sector and put the United States back on track toward economic growth and global dominance.

That agenda will include everything from rebuilding the country's crumbling infrastructure to reorienting our education system with a greater focus on science, math, and engineering. It will include leveraging our manufacturing heft to combat global climate change and build a vibrant clean energy industry. And it will include making our tax code more competitive, and our regulations more streamlined, so that doing business in the United States can be both cheaper and more efficient.

Each item on the agenda will help boost American manufacturing. But no single piece will sufficiently do it all. The U.S. economy requires a comprehensive set of solutions, and so it deserves a truly comprehensive plan. We cannot allow ourselves to be satisfied by the argument that our political system is too broken to make the kind of big changes we need. The U.S. political system has always been partisan. Passing major legislation has always been an uphill climb. But American history is full of moments, critically important moments, where leaders have overcome the obstacles and delivered on essential reforms. This should be one of those moments.

The United States is at a strategic inflection point. Our options are simple—and starkly different. Are we going to fight to compete, as we have throughout our history? Or are we going to stick to assumptions and practices that no longer make sense?

We are still emerging from the wakeup call of the financial collapse. We are still recalibrating, still recovering. Our success will depend on our ability to recognize the problem we face, and our willingness to face it down.

When we had a vibrant, booming manufacturing sector, we enjoyed new wealth and growth unmatched in the world. If we can revive, rebuild, and reinvent that sector, if we can bring back the model that served as the world's greatest force for economic growth, we will enjoy that prosperity once again.

That, I believe, is what we must do. This book is about getting it done.

ANDREW LIVERIS

Acknowledgments

The ideas in this book, as I've tried to make clear, are the product of practical experience. I am lucky to have a job that has taken me around the world, and wherever I go—wherever Dow operates—I am fortunate to work with some of the most brilliant, creative, and hardworking people you could possibly find. Their ideas and actions have an important influence on my opinions and outlooks, so I am pleased to have a chance to thank them publicly.

ACKNOWLEDGMENTS

I am deeply grateful to the women and men of the Dow Chemical Company. I have worked at Dow for more than 30 years, and have always considered it not just a pleasure, but an honor, to work side by side with all of you. Your hard work has yielded countless innovations that have improved the lives of millions of people. Every day that I come to work, I do so with a feeling of great privilege. Serving as your CEO has been the greatest honor of my life.

I also want to thank the people of Midland, Michigan, and of every manufacturing town in the United States. Thank you for keeping your head up, even in tough times, and for believing, as I believe, that the best times for manufacturing still lie ahead of us. You are why I wrote this book.

A CEO, of course, has a great number of responsibilities, and I could not possibly have been able to write this book without the tireless assistance of Matt Davis and Louis Vega. You skillfully managed this project from its inception, and provided strategic advice and input throughout. Dow benefits from your contributions and I truly appreciate all of your hard work and focus toward our success.

Many of the ideas for this book sprang from conversations with two of my most trusted advisers, Declan Kelly and Michael Klein. Thank you both for your insights, your counsel, and your belief that the discussions we began in private ought to move to the public square.

Others played key roles in shaping this book and bringing it into the marketplace of ideas. My literary agent, Raphael Sagalyn, was a vital partner in those efforts from the beginning. And I offer my sincerest thanks to my editor, Pamela van Giessen, and the entire team at Wiley for their collaboration, partnership, and counsel in helping bring this important issue to light.

And most importantly, to my loving and devoted wife, Paula, and our three talented children–Nicholas, Alexandra, and Anthony—thank you for your abiding support during every stage of my Dow career and our ongoing journey through life together. You make what I do worth doing.

A. L.

MAKE IT
IN
AMERICA

Introduction

This isn't just an uncertain world. It's a volatile one.

For years, the United States has been in transition from a manufacturing-based economy to a service-based economy. For years, we entrusted our growth to borrowing and consumerism. For years, we fueled our growth with debt, and with the idea that everything would be just fine.

As we now know, it wasn't.

Too many of us in business and government didn't see it coming. Even if we had concerns—as I did and sometimes shared—none of us imagined just how far, and how fast, we were about to fall.

The financial crisis, the housing market collapse, the ensuing recession and credit crunch—all these have caused no end of pain for individuals and businesses alike. But they aren't the fundamental problem. They're symptoms. The troubles of recent years have unmasked a reality that spent years lying dormant, hidden from public view: the United States no longer has an economic model that's sustainable.

How can this be? Are things really that serious? After all, we are still the largest economy in the world. We still have a larger gross domestic product (GDP) than any country, and higher productivity than any country. And yet, as we reflect on these first decades of globalization, on what they have meant for the United States, we are left with the uneasy notion that we are losing ground we can't gain back.

Our old sense of confidence, of certainty, is at low ebb. That's because we now understand that the free flow of capital in financial markets may create certain benefits, but also produces extraordinary volatility—in raw materials, in final products—requiring American businesses operating globally to contend with challenges on a scale unheard of before, and mostly unacknowledged in America's national conversation.

This volatility has driven entire industries to relocate to the opposite side of the world. It is preventing companies from investing in the United States right when our economy most needs that investment. Recent policy decisions have eased some of the pain, mitigated some of the damage, but it seems no one's talking about a fundamental fix.

Globalization has changed just about everything. In many places it has been a force for progress, but by its very nature, it is also a force of destabilization. It creates opportunities and, at the same time, considerable risk. It gives nations—both developed countries and emerging economies—the ability to prosper in ways we never could have imagined, but also creates new obstacles to growth. Here and elsewhere, globalization has upended old economic models, creating imbalance where order once reigned.

That isn't surprising. Globalization, left to its own devices, favors efficiency. Countries with the capacity for complex financial transactions—like the United States and the United Kingdom—will naturally see the expansion of their financial services sectors. Likewise, countries that manufacture goods cheaply and efficiently are likely to see growth in that sector.

The corollary, which also holds true, is that less efficient sectors tend to erode. This, in turn, tends to exacerbate their inefficiency and hasten their erosion. When manufacturing can be done for less abroad, market forces send it abroad.

That leads to an imbalance within some economies that might seem, at first glance, either inevitable or even desirable. Countries could specialize, the argument might go, and could thereby increase the efficiency of the global economy as a whole. This idea has a certain logical appeal. But it overlooks something fundamentally important about the health of developed and developing economies alike: not all sectors are created equal.

The manufacturing sector, for example, can create jobs and value and growth to a degree that the service sector cannot. As America's service sector expands, and its manufacturing sector contracts, the result isn't a new post-manufacturing economy in which everyone wins. The result, instead, is that certain people win, but the country as a whole experiences massive unemployment. The service sector is certainly capable of generating wealth—as we have seen—but it cannot create the sheer volume of jobs the economy needs in order to sustain a workforce of more than 150 million people.

The places that have been able to succeed (for a time) with highly specialized economies have had, in almost every instance, tiny populations. Dubai, for example, was able to thrive as a service-only economy—at least until the financial collapse—in part because it has a population of just more than 2 million, and therefore needs to create relatively few jobs. The United States does not have that option. There aren't enough high-end service industry jobs in the world to employ the American workforce.

The U.S. economy needs balance across sectors. The nation needs to strengthen its advantages in providing services and intellectual capital for the rest of the world; but we also need to grow things and build things. Only through balance will we find economic strength, stability, and growth over the long term.

Otherwise, if U.S policymakers allow the economy to drift further in the direction of imbalance, the nation will find itself with a highly specialized service sector that supports the few, and a weakened manufacturing sector that can no longer sustain the many. I don't mean to suggest that manufacturing can sustain an entire economy on its own. It can't. But no economy as large as America's—no population as large as America's—can sustain itself without manufacturing. No society can thrive with persistently high unemployment.

There are some who insist that as long as the United States continues to generate the world's greatest innovations, the collapse of manufacturing doesn't matter. Yes, America has been—and should be—the world's greatest innovator. But that alone is not enough. I hope Americans won't buy the notion that they've outgrown manufacturing. Accepting such a future would mean accepting a level of joblessness that would make recent years look like a warm-up.

Passivity is not a growth strategy. For too long, too many smart people have insisted that pure free market principles would help economies (at least the healthy ones) find the right balance. But as we've seen increasingly over the past decade, an economy will not simply balance itself. Doing so requires action; it requires intervention.

A look around the world reveals that the countries that are succeeding economically are not passive believers in free market fate. Instead, they take their future into their own hands and strive for better balance. Brazil, for example, isn't content to be a power in agriculture alone; it is strengthening other sectors as well. And China is working hard to ensure that it is more than the world's factory floor. These nations, like others around the world, are taking action. They are working to fulfill the promise of globalization and to minimize its perils.

Around the world, countries are acting more and more like companies: competing aggressively against one another for business and progress and wealth. Governments are boosting business, creating a climate that attracts and rewards investment, spurs innovation and job creation, and appeals to companies that are less bound by national borders than ever before. Build here, they say, and we'll pay for the land you need. Build here, they say, and we'll cover your workers' salaries for a decade. Build here.

Meanwhile, in the United States, we operate as if little has changed. Our faith in the wisdom of markets may be shaken, but not at a fundamental level—even after the markets have shown they're not always so wise. We assume that because of our greatness, companies will continue to invest here, just as they always have.

We are wrong. Not about America's greatness, but what it entitles us to.

It is time to recognize that if we don't act soon, if we continue to let markets rule in every instance, we will become the global economy's biggest bystander, and potentially, its biggest drain. Our consumers will find themselves with more debt and less money to spend; our businesses will have fewer resources for research and development; our future generations will lack opportunities. It's time for us to recognize the cost of inaction—not just for the United States, but also for the entire global economy.

It's time for us to recognize, whether we like it or not, that for now, in certain key areas, we actually need more government, not less. As CEO of one of the most global U.S.-based corporations, I can tell you, without qualm or question, that I want government more involved. I need it more involved, for the sake of my employees and shareholders. Not as an overzealous regulator, but as a thoughtful partner to thoughtful business—in a shared effort to strengthen our economy.

Risk is inherent in business. I accept that. In fact, I embrace it. There's no upside without a downside.

But any smart company anywhere in the world will do what it can to reduce the uncertainties of doing business, just as individuals do what they can to reduce risk in their lives. Other nations, recognizing this, are working closely with companies to mitigate risks, to create predictability so that businesses can plan and invest for the long term. Together, the public and private sectors set goals, and create mechanisms to ensure those goals are reached.

These countries see what the United States, sooner or later, must realize: it's a false choice to say that you can be either pro-business or pro-government. The old ways of thinking don't apply to the new global economy. Indeed, today, more than ever before, being pro-government is a prerequisite for being pro-business. They must work in concert.

Now, I have no doubt that there will be cynics and skeptics who, upon reading this, will say that I am calling for nothing more than corporate welfare. After all, businesses that espouse free market principles are also known to lobby governments for policies that they believe will improve the environment for their investments. It's no secret that Dow, too, has lobbyists, and we put our policy objectives on our web site for all to see.

I don't suppose there's anything I can say to win over the skeptics. But I hope that fair-minded readers—who share my strong belief that the best engine for economic growth is a healthy, strong private sector—will see that the case I make in this book transcends the particular interests of Dow, the chemical industry, or even the manufacturing sector. I believe that everyone in America has a vital stake in whether businesses are willing to stay and grow within these borders.

At a time when U.S. companies—run by patriotic people—are moving offshore at the fastest rate in history, we should, at a minimum, recognize that the model we are relying on isn't working. When those companies leave, they take jobs and growth—and all the prosperity that follows from them—offshore, too.

The United States must do more to reverse the tide.

By "the United States," I mean all of us—the public and private sectors. We must build a new partnership for prosperity. We must draw on America's rich history of collaboration between business and government—a relationship that, despite its obvious tensions, has brought immeasurable benefits to the people of this country and the world. In World War II, companies worked with government to create, as Franklin D. Roosevelt called it, "an arsenal for democracy." Later, government investments in microprocessors, for example, made those innovations scalable, and affordable to the mass market.

Countless other innovations were the product of government-funded research and development. More recently, government action has helped turn the auto industry from bankruptcy to profitability in less than two years. Government intervention stabilized the financial markets and unfroze credit lines.

Government, of course, often stumbles, oversteps, and makes terrible decisions. Every business person I know sometimes wants government to get out of the way. But that doesn't mean it should get out of the picture.

I am a businessman, and I am unabashed in my boosting of business. I believe deeply in entrepreneurship, in the power of capitalism to improve lives. But I do not agree with those in the business community who think that our future success depends on government shriveling up and dying on the vine.

We cannot afford to get stuck in the tired old debate between pure free-market philosophy and state socialism—as if those are the only two economic models from which to choose. We should not assume that the rules of the twentieth century economy apply to the twenty-first century. Globalization has truly rewritten the rules. The new economic realities apply broadly—not just to the United States. There isn't a major power in the world that can succeed over the next several generations without figuring out how to balance or re-balance its economy.

And we need them to succeed. In this era of interdependence, we need more people in more nations to succeed. There are times, of course, in business, in global competition, where we're playing a zero-sum game. I want to win, and, let's be honest, I want you to lose. In a similar sense, nations want to maximize their global market share; they, too, want to win. But the calculus we apply has to be different than it is in business, because our economies are more integrated today than ever before.

In an era where one country's economic troubles—whether that country is as big as the United States or as small as Greece—can set off a global chain reaction, we all have a stake in each other's strategic decisions. And that's not necessarily a bad thing. Indeed, I believe the enduring reality of globalization can be the rising tide—one that truly lifts all boats.

As the world's largest economy, the United States therefore has an obligation not just to its own people, but to the people of the world, to get this right. To no longer accept the shuttering of factories and offshoring of jobs as inevitable. To reject once and for all the idea that manufacturing is somehow optional, or incidental to our future.

On the contrary: manufacturing *is* America's future. Not just its past. Manufacturing is the foundation upon which our economic prosperity, our growth and wealth and jobs depend. At the center of our economic problems lies the hole that was left when manufacturing started to disappear. And at the center of our solution is a strategy to rebuild that once vibrant sector.

Chapter 1

The Rise and Fall

In 2007, in a secretive Silicon Valley research facility known only as Lab 126, engineers and designers successfully developed a new product with the potential to revolutionize the way we read books like this one. It was called the Kindle, and it would represent Amazon's first attempt at selling a product of its own.

What was special about the device was not just that it was a portable library of books, but that it used an innovative electronic ink. Unlike the pixels of a standard computer screen, the tiny capsules of electronic ink change what appears on the screen without illuminating it, allowing the Kindle to simulate a printed page and to be read even in direct sunlight.

The team at Amazon was excited about the potential of their product and began to search for a U.S. company that could manufacture it. To produce the special ink beads, Amazon partnered with a Massachusetts-based company, appropriately named E-Ink. The company had been started by researchers working at the MIT Media Lab and was now one of the only manufacturers in the country capable of producing electronic ink devices.

But E-Ink did not have the technology to build the Kindle screen itself. For Amazon to build the entire product, the company would have to partner with an additional manufacturer.

That search began in the United States; but it did not end there. The production technology required to build the screen was similar to what's used to make LCD televisions. Amazon needed a manufacturer with experience in that field. But there was a problem: Amazon couldn't find one in the United States. Though the LCD television was originally the product of American research and development (R&D), the entire industry had been ceded to Asia in the 1990s, when Asian countries offered U.S. television manufacturers a business environment too attractive to resist. By 1995, not a single LCD panel was being manufactured in the United States.*

As a result, Amazon was forced to look overseas to find a manufacturer with the expertise and capability to make the Kindle screen. Eventually, Amazon turned to a Taiwan-based company. The irony was a painful one for U.S.-based Amazon. Though its innovative new product had been developed in America, it would

*Olevia started manufacturing LCD television in the United States in 2006. But according to *Popular Mechanics*, the panels themselves are manufactured in Asia, and are only assembled in the United States.

not—could not—be built in America. Today, as Harvard Business School professor Willy Shih has said, when a Kindle is purchased by an American consumer, it adds to our trade deficit.

Not long after Kindle production began, the Taiwanese manufacturer, Prime View, realized that it could make the Kindle at a lower cost if an ocean wasn't separating the building of the screens from the creation of the ink. As a result, Prime View bought E-Ink and moved its headquarters—and the electronic ink industry—to Taiwan.

The story of the Kindle is not unique. On the contrary, it is becoming a defining experience of the U.S. manufacturing sector.

Over the past several decades, the United States has watched entire industries disappear from its shores—only to reappear abroad. Industries from solar panel technology to highly advanced computer circuitry, from wind turbines to smart phones—industries that were born in the United States—now exist predominantly elsewhere. Between 2001 and 2010, U.S. companies were forced to shutter more than 42,000 factories. A third of all manufacturing jobs—a full 5.5 million—have disappeared. The entire sector is hemorrhaging.

Today, instead of manufacturing making up 28 percent of GDP, as it did in the 1950s, it makes up just 11.5 percent. Instead of exporting billions more than we import, the United States now faces a half-trillion dollar trade deficit.

As Richard McCormack, a leading thinker on manufacturing issues, has noted in *The American Prospect*, the furniture industry lost at least 60 percent of its production capacity in the United States between 2000 and 2008. By 2009, the U.S. auto industry was in shambles, requiring a federal bailout to survive. The Chinese are now the global leader in auto manufacturing; in 2008, they made half a million more cars than the United States. General

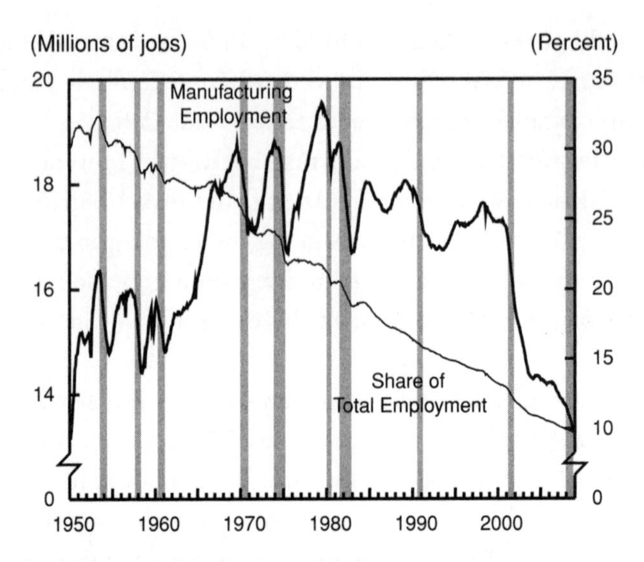

Manufacturing Employment Since 1950
SOURCE: Congressional Budget Office: Factors Underlying the Decline in Manufacturing Employment Since 2000.
DATA SOURCE: Based on data from Department of Labor, Bureau of Labor Statistics.

Motors, once America's largest manufacturer, now plans to build cars in China and export them back to the United States. Wages in textiles, textile products, and apparel are expected to be cut in half by 2018.

Perhaps the most telling statistic of all: In 2008, 1.2 billion cell phones were sold worldwide. Not a single one was built in the United States.

How We Fell Out of Love with Manufacturing

How did a nation that once defined and distinguished itself by the things it built, a nation whose economic engines were driven by manufacturing, end up ceding its identity—and its future—to other nations?

The United States used to be the world's greatest manufacturer. After World War I, U.S. manufacturing ability made the country an economic leader among nations. After the Great Depression, it was manufacturing—specifically, the manufacturing of war materiel—that led the United States out of the depths of economic despair. In 1953, General Motors alone generated 3 percent of U.S. gross national product. For the 30 years after World War II, America experienced a post-war boom driven by manufacturing that helped build a vibrant middle class, not just in the United States, but around the world. Between 1947 and 1973, family incomes doubled. According to the State Department, gross national product grew 50 percent between 1940 and 1950 and another 67 percent between 1950 and 1960.

From the time of America's founding, manufacturing growth defined the nation's economic strength. That changed in the mid-1970s. In 1975, we were still exporting more goods than we were importing. But that was the last time we did so. In the 35 years since, the United States has consistently been burdened with a trade deficit.

Other major world powers had fully rebuilt their manufacturing bases after World War II, thanks largely to the Marshall Plan. That led to increased competition abroad. The United States signed onto multiple free trade agreements, reducing the barriers to importation. Cheap products from China and India, from Taiwan and Brazil, began flooding the U.S. marketplace.

Through the 1980s and 1990s, the United States continued to manufacture more goods than any other country. But competition from developed and developing countries alike was putting a strain on manufacturers across the board. By the 1990s, the fall

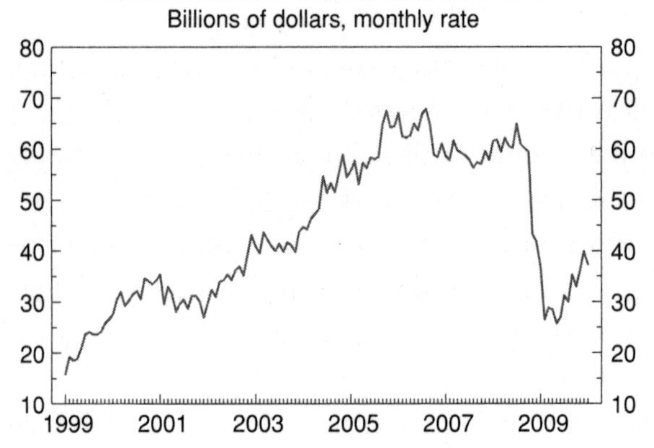

*Note: Services totals include revisions to the December 2009 data.

Trade Deficit

SOURCE: International Trade Administration: U.S. Export Fact Sheet, January 2010.

of Communism ushered in a new era of free markets. Even China, which many believed might keep its economy closed, turned its engines full-tilt toward capitalism of a distinctly Chinese variety. In 2001, China joined the World Trade Organization. By opening up its economy and breaking down major trade barriers, the Chinese could rival the United States as the highest-producing manufacturing nation in the world.

The rise of the Internet, too, marked a major turning point for global manufacturing. It allowed businesses to communicate and partner more easily with foreign suppliers. It made the world smaller, more interconnected. It lowered the barriers of entry into emerging markets and made it cheaper to operate a far-reaching global supply chain.

For many businesses, these opportunities brought levels of success previously unimaginable. As companies entered emerging global markets, they found not just those who could produce their goods, but those who could consume them.

This free flow of global capital gave businesses a greater choice of where to build a new factory or plant. It encouraged countries to aggressively compete for foreign investment, knowing that if they could attract new business, they could build their economies. In addition to offering cheap labor, countries offered businesses lower tax rates and easier-to-navigate regulatory regimes, among other incentives. When manufacturing companies assessed their financial position, it started to make more sense for them to move jobs offshore than to continue to operate in the United States.

And so they did. And today, as a result, the entire U.S. manufacturing sector is in crisis. The scaling process—the process of taking an idea out of the lab to mass production—is barely happening here anymore. Ideas born here are getting built elsewhere.

At the same time, America shifted its focus from manufacturing to the service sector. In some sense, this was to be expected. Our current generation of business and political leaders grew up, for the most part, in the post-war period. This was a time not only of economic stability, but of sustained prosperity. Many of the children of that time period had parents who worked on the assembly line—and hoped for a life for their children outside the factory. Perhaps a white-collar job, a position in management. Or a professional degree and a career as a lawyer, a doctor, a banker. This was the great era of upward social mobility—always a proud feature of American life, and never more so than in the boom years of the 1950s and 1960s. Indeed, the prosperity of that period allowed millions of young Americans to get a better education than their parents, and, ultimately, to get a better paying, higher valued job.

This was all a uniquely American success story. But it had unforeseen consequences, some of which are only now becoming clear. These consequences were cultural and economic. What happened was that somewhere along the way, a leadership class emerged—in the public and private sector—that no longer valued the manufacturing sector. Yes, they were grateful for the sacrifices their parents had made on their behalf—but they had come to see manufacturing as merely a phase of economic development, something that a nation eventually outgrows, just as they had. So we populated the corporate world with service professionals, and filled Congress and the White House with lawyers and lobbyists, leaving manufacturers on the margins of our public dialogue.

As the manufacturing sector slowly sunk into a sustained crisis, there weren't enough people in positions of leadership who truly understood its importance, who truly appreciated the risks we faced if we let the sector erode.

What ultimately happened to the manufacturing base in the United States, then, was not the product of fate. It was the product of choices we made as a nation.

The Multiplier Effect

Most politicians and business leaders acknowledge that manufacturing is an important part of the economy. They travel the country, particularly traditional manufacturing areas, and talk in front of crowds at manufacturing outfits. Let us assume that these people are sincere when they proclaim their fidelity to manufacturing and manufacturing jobs. However, what they fail to recognize is that the loss of American manufacturing is going to be felt far beyond the streets of working-class Rust Belt communities. It is going to affect far more than the families who have lost jobs, and the towns that have lost entire companies. It's going to affect almost everyone. Because without manufacturing, the U.S. economy cannot—and will not—sufficiently grow.

Let me explain why:

Manufacturing, more than any other sector, creates jobs outside its own sector. These jobs range from construction and mining to jobs in fields like packaging and telecommunications. Even in 2009, a year when manufacturing was experiencing its sharpest decline to date, the sector still supported nearly 7 million nonmanufacturing jobs—jobs outside the plant. Those are jobs along an extensive supply chain.

We call this the multiplier effect. A new manufacturing facility will create demand for raw materials, construction, energy, supplies, and services. According to the U.S. Bureau of Economic Analysis, of all sectors, manufacturing has the biggest multiplier effect. As the Manufacturing Institute notes, "every dollar in final sales of manufactured products supports $1.40 in output from other sectors of the economy." Compare that to the service sector, which, according to the National Association of Manufacturers, supports just half as

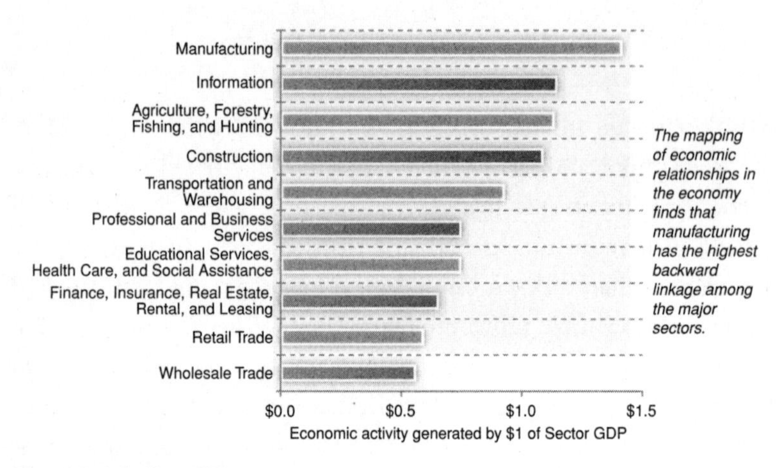

The Multiplier Effect

SOURCE: Copyright © 2009 by The Manufacturing Institute. *The Facts about Modern Manufacturing, 8th ed.* 2009.

DATA SOURCE: U.S. Bureau of Economic Analysis, 2007 Annual Input-Output Tables.

much—$0.71—in output for every dollar in final sales. These are high paying jobs that we lose every time we move a plant offshore.

Consider the iPhone. Open the box of a brand new iPhone and you will read this proud statement: "Designed by Apple in California." Now look at the back of the device and you will find, in letters so small they are difficult to read, this admission: "Assembled in China." That's a story we know all too well.

But what if, instead, the back of the iPhone read, "Designed *and* Assembled in California"? What would be the difference for the U.S. economy? If Apple were to open their factory next to their R&D facilities in the Silicon Valley, instead of Shenzhen, it would create jobs in California for an entire supply chain.

Businesses from packaging and office equipment to telecommunication services and building maintenance would have reason to hire new workers. Companies that mine raw materials would have reason to increase production. Local restaurants would have new customers. Local malls would have new patrons. The plant would become a new economic engine, leading to thousands of new jobs in town. The iPhone would create a mini-economy right here at home. A few hundred jobs inside the factory could mean a few thousand jobs outside of it.

Open that facility in China—or any other country—instead, and all those jobs, all that opportunity and prosperity, evaporate. New jobs and a new supply chain will emerge around the Chinese factory, and the wealth that could have been added to our economy will be added to theirs. Apple shareholders still see their returns. The engineers in California still get their paychecks. But the fruits of their labor accrue elsewhere. Indeed, for every Apple worker in America, there are 10 Apple workers in China.

But that isn't the only reason that manufacturing is a prerequisite for the level of economic growth we need. Manufacturing also adds value to the economy in a way and at a higher rate than other sectors. I'll explain this further in Chapter 2.

Manufacturing is also the primary driver of research and development, generating new ideas, inventions, and intellectual property that fuel long-term growth. More than two-thirds of the money spent on R&D in the United States is spent in manufacturing. Without a vibrant manufacturing sector, that R&D will be done not by the United States, but by its major competitors. Over time, that will leave America dependent on intellectual property that's created by other countries; America's ability to generate its own growth will atrophy. This, too, will be explained further in the next chapter.

The United States, therefore, cannot sustain the level of growth it needs—and has come to expect—without a stronger manufacturing sector. That presents a serious challenge. The United

States, like any nation, needs its economy to constantly grow. Every day the population increases. Every day new workers join the workforce—or seek to do so. According to the Economic Policy Institute, the United States has to create more than 125,000 new jobs every month just to keep pace with population growth. As we grow, the economy must grow with us.

Healthy economic growth results in higher wages, higher profits, a stronger middle class, and a better standard of living. It is the engine by which we prosper. Without economic growth, wages will stagnate, as will profits. The unemployment rate will stay high. Reductions in standard of living will be felt across the board. Without economic growth, the American dream, as we've imagined it, will come to an end.

Manufacturing Tomorrow

Designing here and building there is sometimes smart, sometimes necessary. But in general, it is simply not a formula for long-term economic success. If we want to create jobs in the United States—as we surely do—we cannot ignore manufacturing. We must revive it.

Yet, if this problem is so critical to the economy—and it is—it begs the question: Why haven't our political leaders focused on it? Sure, we hear lip service when presidential candidates are campaigning in Youngstown, Ohio, or Allentown, Pennsylvania, but why haven't concrete actions followed those words?

I believe part of that problem is a fundamental misunderstanding of what manufacturing is, and what a new manufacturing sector in America would be.

Too often in our political dialogue, when people talk about manufacturing, they talk about it in the wrong way:

"Are jobs that were a relic of an earlier time ever going to return? Are the jobs that have gone overseas worth bringing back?"

We heard this conversation a lot during the 2008 presidential election. While traveling through Michigan in January 2008, former Massachusetts Governor Mitt Romney told voters, "I hear people say, 'It's gone, those jobs are gone, transportation's gone, it's not coming back.' I'm going to fight for every single job. I'm going to rebuild the industry."

Senator John McCain, in response: "I've got to give you some straight talk. Some of those jobs that have left the state of Michigan are not coming back. They are not. And I am sorry to tell you that."

This debate misses the mark. Too often, the players in these debates are either arguing that we should revive the manufacturing jobs of the past, or instead, that we ought to write off the entire sector. But I'm not arguing that we should bring the jobs of the 1930s, 1940s, or 1950s back to our country. I'm not interested in restoring the manufacturing sector of old. I'm arguing that we should build a different kind of sector, an advanced manufacturing sector, one that offers high paying jobs in high-tech, state-of-the-art industries. Industries of the future. Industries that are changing the world. Industries that are changing the way we live in it.

When we create advanced microprocessors that can double the speed of the fastest computers, that's manufacturing. When we construct wind turbines that can power an entire city without any carbon emissions, that's manufacturing. When we create chemicals that can dramatically improve the life of a hybrid battery, that's manufacturing. When we build complex robotics that can assist a neurosurgeon in removing a brain tumor, that too is manufacturing.

It's not the industries of the past I'm worried about losing. It's the industries of the future.

And they're already disappearing.

For more than 30 years, for example, the U.S. government funded research and development labs that produced some of the most important breakthroughs in solar energy. According to the Department of Energy, the United States spent $15.4 billion (in 2008 dollars) between 1978 and 2007 on renewable energy R&D. But it was Japan—not the United States—that commercialized those panels. And now it's China—not the United States—that dominates the solar manufacturing industry globally. That massive national investment of $15.4 billion has created a U.S. solar manufacturing industry that today employs just 10,000 people.

As Intel founder Andy Grove wrote in *BusinessWeek*, "Today, manufacturing employment in the U.S. computer industry is about 166,000 lower than it was before the first PC, the MITS Altair 2800, was assembled in 1975. Meanwhile, a very effective computer manufacturing industry has emerged in Asia, employing about 1.5 million workers." He also notes that the biggest of these Chinese companies, Foxconn, saw its revenues in 2009 hit $62 billion, "larger than Apple, Microsoft, Dell, or Intel. Foxconn employs over 800,000 people, more than the combined worldwide head count of Apple, Dell, Microsoft, Hewlett-Packard, Intel, and Sony."

By ceding these industries to other countries, we aren't just losing out on today's manufacturing jobs. We're losing out on the production of tomorrow's innovations, on the progeny of the products being built today.

Some disagree. Princeton University economist Alan Blinder sees only upside to our loss of the television market, for example. "The TV manufacturing industry really started here, and at one point employed many workers," he wrote in his book, *Offshoring of American Jobs*. "But as TV sets became 'just a commodity,' their production moved offshore to locations with much lower wages.

And nowadays the number of televisions sets manufactured in the U.S. is zero. A failure? No, a success."

But it wasn't just televisions we lost. As Andy Grove notes, "Not only did we lose an untold number of jobs, we broke the chain of experience that is so important in technological evolution. As happened with batteries, abandoning today's 'commodity' manufacturing can lock you out of tomorrow's emerging industry."

This, as I described earlier, is exactly what happened with the Kindle. It may have been the previous generation of innovations—LCD televisions—that the United States gave up on producing. But it was the next innovation—the Kindle—that the United States, as a result, wasn't able to build.

It's that dynamic that could deliver a death blow to U.S. manufacturing. As other countries master next-generation manufacturing techniques, and as they gain expertise in innovation, the United States could find itself falling behind for good, and out of a game whose rules America used to write.

Surviving the Crisis

The companies that have survived have been those that were able to transform, to adapt their business models to match the changes in the global economy. At Dow, that is a story we know well. With new global markets and high volatility in basic resources, for Dow to continue to thrive, it had to change.

Through Dow's growth from a small business in Midland, Michigan, to a major multinational corporation, we focused primarily on producing basic chemicals. We were known as "the chemical company's chemical company." Most of the products we sold were chemicals and plastics that would be used by other chemical manufacturers to produce even more advanced chemicals.

But in this new age, it made little sense—from a financial perspective—for us to continue our focus just on a basic chemicals portfolio. At Dow we recognized that we needed to transform ourselves, reorient our mission, our plan, and our projects toward products that yielded higher value and higher demand.

Today, that transformation is nearly complete. Two-thirds of our businesses are high-margin, high-growth sectors such as advanced materials, specialty chemicals, performance products, and agricultural sciences. That makes us more competitive, more innovative, and better-positioned for sustained growth over the coming decades.

That's the same kind of transformation I envision for the U.S. manufacturing sector.

I am passionate about manufacturing. I am passionate about these kinds of new, advanced products, about the promise they offer not just to our economic well-being, but to the quality of our lives.

Take one of Dow's latest innovations, the POWERHOUSE solar shingle, for example. In 2009, Dow developed a solar cell so flexible and durable that it could be installed as a shingle on an ordinary roof. This innovation is unique to Dow, and a true game-changer. Today, there are scores of customers who would like to run their homes on solar energy, but they can't, either because it's too expensive, unavailable in their area, or, as is often the case, their homeowners' association won't allow it. No longer. These solar shingles can be installed by any contractor on any home, just about anywhere. I'm proud to report that *Time* magazine named our solar shingles one of the "50 Best Inventions of 2009."

And that's just one example. Here's another one—computer chips. Manufacturers are constantly working to make faster chips,

chips that process more information at less cost. To make these powerful new integrated circuits, layers of chemicals must be applied to them. But there is no margin for error. If one layer is off by just 10 nanometers—one ten-thousandth of the width of a human hair—it can ruin the entire circuit.

When you measure success in terms that small, you're working on the molecular level. So if you're a chip fabricator, you need the right molecule, and you need it to behave the right way. That's our specialty. You wouldn't trust that level of quality to any manufacturer. Performance trumps cost. And it's these kinds of high-performance products that can revive the manufacturing sector. As a nation, we have to evolve from manufacturing only the basics to manufacturing the most advanced products. Companies do this to compete. Nations must, too.

A Tale of Two Nations

The United States cannot afford to fail. To see the potential consequences, we need to look no further than our ally across the ocean. The United Kingdom's manufacturing sector is in even more severe decline than the United States. The British economy was once one of the most productive in the world, with a vibrant manufacturing sector that ushered in substantial economic growth. But even before the most recent recession, the UK manufacturing sector had been lagging behind the rest of the world. Between 1979 and 1982, its output dropped 18 percent. Between 1997 and 2006, a period in which U.S. output grew by 30 percent, British output declined. In the early 1980s manufacturing made up 31 percent of GDP. By 2008, it was just 13 percent. Think of it this way: In 1980, one in four of all UK jobs were in manufacturing. By 2008, that number was just one in ten. Then the recession hit. Between December 2008 and February 2009, British manufacturing showed its steepest decline since records began being kept.

The consequences of their manufacturing weakness are evident. It is no surprise that the United Kingdom was the last of the major economies to finally come out of the recession. A country without a strong manufacturing base is without the economic tools to spark sustained recovery. Even when consumer spending increases, the goods they are buying are not produced at home. As a result, it's not the British economy their spending is stimulating. It's foreign economies. The United Kingdom is overwhelmed with debt, and in the coming years must make incredibly difficult financial choices. But without manufacturing, the growth they'll need to overcome their challenges will continue to be far out of reach.

This is a bad situation not just for the British, but for the world. When the sixth largest economy finds itself saddled with an economic crisis it cannot solve, the consequences will be felt worldwide.

Here's what we know with certainty: nations that support manufacturing can—and do—thrive. Germany is an extraordinary example of a country with the correct economic priorities. As journalist and manufacturing expert Eamonn Fingleton notes, from 1998 to 2008 Germany went from a trade deficit of $5.9 billion to a trade surplus of $267.1 billion. In that same period of time, the U.S. trade deficit more than doubled, from $224 billion in 1998 to $569 billion in 2008. Even in the depths of recession, the German economy was stable enough to keep unemployment well below 10 percent. To put their manufacturing might in perspective, Germany makes up just 1.2 percent of the world's population, but German industry accounts for 17 percent of global market share.

What's the difference? What is Germany doing that the United Kingdom isn't? According to Fingleton, "The secret to the German system's success is, in large part, a strong national commitment to advanced manufacturing." The German government has a keen sense of the importance of manufacturing, and has made investments to support the sector, even as they transition their economy. That's why manufacturing makes up 20 percent of the German economy, but only 11 percent of the U.S. economy. And it's why, in the race for a competitive long-term future, Germany is far ahead of the pack.

We can be like the United Kingdom. Or we can be like Germany. And it's entirely up to us.

By now it's probably easy to tell that I love manufacturing. I know it has the power to be a transformative force. I come to this topic with a unique perspective. As CEO of one of America's five most global corporations, I have seen firsthand the best practices in place in other countries. We have much to learn from their examples. Because we are one of the world's biggest suppliers to other manufacturing companies, I am also intimately familiar with the plight manufacturers face. They are our customers. And because we are the world's biggest chemical manufacturer, I know personally what needs to be done to keep businesses like ours growing and prospering.

Globally, manufacturing isn't dying. It's evolving. Our opportunity for long-term economic growth depends on our ability to evolve with it.

We know that with the right policies and the right business decisions, manufacturing can work in the United States. Dow just recently opened new plants in Midland. We know that building plants here is still a possibility, and more importantly, we know that companies like Dow, which have moved jobs overseas, still want to see manufacturing happening here at home.

We have a choice. We can either continue down the path we're on, one that is leading us the way of weakened economic powers. Or we can make decisions and choices today about our future, ones that put us on a path toward new economic prosperity. Failure is not inevitable. But to avoid it, we must act now.

Chapter 2

Separating What Can't Be Separated

How can it be that a crisis this serious is being so largely ignored in the United States? Even in a country where inaction and gridlock tend to define our political system, you would think—or at least hope—that when our future economic prosperity is on the line, when our opportunity to remain the world's leading economy is at risk, at least some in Washington would do something about it.

Yet for the most part, that has not happened. While manufacturing jobs continue to disappear from U.S. factory towns, politicians seem more concerned with providing transitional relief—job training and unemployment benefits—than they do with fixing the fundamentals that are causing the problem in the first place.

Part of that is just the nature of politics, of politicians who are obsessively focused not on the long-term future of the country, but on the short-term future of their own job security. Why focus on a major, long-term problem without an easy solution when you can focus on a minor, short-term problem that will help with an upcoming reelection? But while that frustrating dynamic plays a role in this and other issues, it's not the only reason that politicians have mostly ignored the manufacturing crisis. It's not even the primary reason. Politicians are ignoring the problem because they've been convinced that it's not actually a problem.

Over the past 10 years, as the country has hemorrhaged manufacturing jobs at a rate and consistency unmatched in its history, there has been plenty of reason for alarm. But a conventional wisdom emerged from among prominent economists and policy experts who argue that the manufacturing sector isn't actually in decline and that the changes we are experiencing today will be good for the United States and the rest of the world, even if they cause temporary pain.

Gregory Tassey of the National Institute of Standards and Technology refers to those who refuse to accept the existence of a manufacturing crisis as "apostles of denial."

"Unfortunately," writes Tassey, "while trends indicating declining competitive positions have been identified and proclaimed by an increasing number of analysts, they nevertheless continue to be rejected or minimized by an even larger number of other analysts and policy makers." And that group is quite bipartisan:

On the right, Dan Ikenson of the Cato Institute: "U.S. manufacturing is not in decline . . . In absolute terms, the value of U.S. manufacturing has been growing continuously, with brief hiccups experienced during recessions, over the past several decades." He continues, "As a percentage of our total economy, the value of manufacturing peaked in 1953 and has been declining since, but that is the product of rapid growth in the services sector and [is] not . . . an indication of manufacturing decline."

On the left, Matt Yglesias of the Center for American Progress wrote a similar sentiment: "I do feel compelled to point out that the 'decline' of American manufacturing is very frequently overstated. You often hear it said that America 'doesn't make things anymore' but it's just not true . . . our manufacturing has [just] gotten much, much, much more efficient."

One of the most prominent among those "apostles" is economist Robert Reich, who served as Secretary of Labor under President Bill Clinton and who has written prolifically about the transforming economy. In 2009, Reich wrote an article for *Forbes* with a thesis that is as close to the opposite of the argument of this book as one could possibly write: "It doesn't make sense for America to try to enlarge manufacturing as a portion of the economy."

I am an admirer of Reich's. He did a lot of important things for the American middle class while serving as Labor Secretary, and has committed a life of scholarly work to improving economic conditions for working Americans. But, on the issue of manufacturing's importance, he and I part ways. Reich and others argue that the cause of manufacturing job losses is not that plants and factories are moving offshore; it's that plants and factories are getting better at making the same products with fewer workers. "Even if the U.S. were to seal its borders and bar any manufactured goods from coming abroad," he argues, "we'd still be losing manufacturing jobs. That's mainly because of technology."

Reich thinks that the reason we have seen so many manufacturing jobs disappear from the sector is that we have made dramatic improvements in the productivity of our factories, replacing workers with robots, replacing complex tasks with computers. Gregg Easterbrook, author of *Sonic Boom*, a book about the extraordinary speed of globalization, makes a similar argument: "It's a misconception that the new era of global competition has caused some kind of jobs wipeout in the United States . . . this would have happened regardless of globalization, or silicon chips, or offshoring, for that matter, regardless of whether Toyota had ever come into existence."

Easterbrook cites an oft-quoted study by Robert Lawrence of Harvard University and Martin Bailey of McKinsey & Company that argues that 90 percent of manufacturing job losses were the results of productivity gains. "Trends like this," Easterbrook concludes, "are really bad for the factory workers whose jobs are lost—but good for everyone else. Prices decline, goods rise in quality, and money is freed up for the many sectors of society where productivity increases are neither practical nor necessarily desirable."

It sounds like a pretty compelling argument, I'll admit: We're not losing manufacturing jobs because of China and India and Mexico and Brazil. We're losing them because we are getting better at manufacturing.

And there's more. Because we are getting more productive, the argument goes, the fact that we are losing manufacturing jobs doesn't mean we are losing manufacturing output. In fact, as Reich will tell you, we are actually producing more goods now than we have in our entire history. Even in the past 10 years, as manufacturing jobs have disappeared, our total output has actually increased.

So you see, they conclude, there is nothing to be worried about. We are still producing a record amount of goods, and that's what adds real value to the economy. Sure it's painful for the folks working in manufacturing towns, for those who lost their jobs because the plant they spent a career at shut down, but the fundamentals of the U.S. economy are strong, and we'll be able to prosper in the future, even if manufacturing continues to make up a smaller and smaller fraction of our total GDP.

That's the first part of their argument: the manufacturing sector is actually doing just fine.

The second part of their argument is that we can continue to prosper, even without a vibrant manufacturing sector, as long as we focus on creating service sector jobs—doctors, lawyers, journalists, engineers, consultants—jobs Reich refers to as "symbolic analysts." He argues that as these positions overtake manufacturing jobs as a proportion of the economy, the losses in one sector will simply be made up for by gains in the other.

"The nations with the highest percentage of their working populations doing symbolic-analytic tasks will have the highest standard of living and be the most competitive internationally," says Reich. As long as we continue to be the world's great innovators, he argues, as long as we have the smartest engineers and inventors, designers and developers, we will prosper. If the ideas are ours, he concludes, it's okay if the manufacturing is theirs.

Even Tom Friedman, who I greatly respect, has implied a similar point. In describing China's dramatic move to increase their green energy capacity in a January 2010 column, Friedman explained that "even Chinese experts will tell you that it will all happen faster and more effectively if China and America work together—with the U.S. specializing in energy research and innovation, at which China is still weak . . . with China specializing in mass production."

It really does sound like a great argument. It's compelling; it's persuasive. And it has the capacity to offer enormous relief. We are facing a crisis that seems difficult, if not impossible to stop. What better way to address it than to conclude that it's not actually a crisis at all? What better way than to conclude that our loss of manufacturing jobs isn't the result of something we're doing wrong? It's just part of a healthy economic evolution, one that puts us on pace to continue to do great things—just different things. We innovate. They produce. When an answer to a problem is that simple, it's no wonder it gets the attention of so many political leaders.

I wish it were accurate. I really do. It would be fantastic news for the American people and it would continue the rather idealistic belief that globalization doesn't create winners and losers—it just creates lots and lots of winners. I would sleep easier at night if I believed what they are selling.

But I can't. Because what they are selling is just plain wrong. They are wrong about what's causing manufacturing jobs to disappear. And they are wrong about what its effect has been and will be down the line.

The Truth about the Manufacturing Crisis

Has the United States enjoyed increased productivity? Absolutely. Productivity gains have been a key feature of the manufacturing sector since Henry Ford first created the assembly line. We get better at building things; we introduce new machines and new processes that make the production of goods faster and cheaper. It used to take GM many days to assemble a single vehicle, for example. Now it takes just over 22 hours.

We have plenty of factories and plants around the country that need far fewer workers operating them than they used to. And because we have sought to increase our productivity relentlessly over the past century, we can confidently say that the United States has by far the most productive factories and the most productive workers in the world.

But if productivity has always been a part of manufacturing's success, why would gains in productivity today cause the relatively sudden loss of so many jobs? There's really only one way that would make sense: if our recent gains in productivity were dramatically bigger than they had ever been before.

You can imagine a scenario in the extreme, for example, where all factories in the country are taken over by highly efficient robots, such that we continue to produce the same amount of goods without ever needing another manufacturing job again. In that situation, our output would still be high, but essentially all manufacturing jobs would have dried up. You wouldn't need something quite that dramatic to explain why in the past 10 years so many millions of jobs have completely disappeared. But you'd need to see some pretty substantial productivity gains over the past decade for this argument to be plausible.

But that hasn't happened. In fact, it hasn't come close.

Let's take a closer look: Between 1989 and 2000, manufacturing employment in this country was relatively stable. And during that period of time, U.S. manufacturing showed an increase in productivity of 3.8 percent each year.

Now between 2000 and 2007—a time period that completely excludes any potential losses from the Great Recession—the United States had almost the exact same productivity gains year over year—3.7 percent. Yet 3.5 million jobs disappeared in those years.

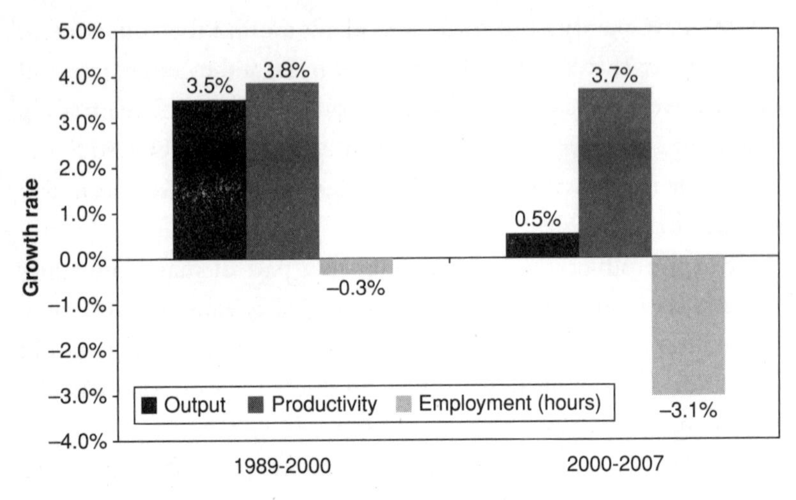

The Truth about Manufacturing: Productivity Isn't Causing Job Losses

FROM: http://www.epi.org/analysis_and_opinion/entry/china_trade_and_jobs-responding_to_myths_and_critics.

DATA SOURCE: BLS data.

How could it be that over the first time period I mentioned, productivity gains didn't cause major job losses, but over the second period, where productivity was essentially unchanged, huge job losses occurred?

The reason is that those job losses have very little to do with productivity. They have to do with aggressive international competition. They are primarily the result of other countries attracting American business to their shores.

During that first time period, 1989–2000, our output grew by 3.5 percent a year. We were getting more productive, but we were also producing a lot more each year. But between 2000 and 2007, output growth completely cratered, falling to just 0.5 percent. That's an 85 percent decline.

Now Reich, Easterbrook, and the other crisis-deniers seem to be saying that we ought not be worried about manufacturing because we are producing more now than we ever have before. They have their facts right, but their conclusion wrong. It's true that we are producing more now than we ever had before in absolute terms. But that's not something you should be particularly impressed by. Are we making more than we did in 1950 when things were booming? Of course we are. There were only 152 million people in the United States then. There are about 310 million now. Not to mention how many more global consumers there are now than there were then. That's the whole point of this issue: our population is growing every day and we need our economy to grow at a healthy rate to keep up with it.

In economics we don't measure success in absolute numbers. We measure success in growth. You cannot grow your economy unless you grow your economic output. So to say that we shouldn't be worried about output because the absolute total number of products we made last year is more than the amount we made the year before completely and totally misses the point. It's not a question of whether we are producing more each year. It's a question of how much more. Our economy was booming when we were producing 3.5 percent more year over year. But it will come to a near halt if we continue to only increase our output at our current pace.

Those who deny that manufacturing is in crisis have the cause wrong: job losses in manufacturing are not the result of productivity gains. The have the effect wrong: we aren't actually producing enough goods at home. And they have the conclusion wrong: No, everything will not turn out just fine.

We know this. We see it every day. We know that more than 42,000 factories have been shuttered in this country in the past decade. We know that some of them have closed because the companies that own them can no longer afford to produce goods in the United States and still turn a profit. We know that some that have closed have done so because they cannot compete with foreign manufacturers who are able to produce the same products for far less. And we know that thousands of the shuttered factories have reopened abroad.

In 2008, Sony announced that it would close its last LCD television factory in the United States, and serve the U.S. market, instead, from its plant in Mexico. In 2009, Dell Inc. closed its computer plant in Winston-Salem, North Carolina, laying off almost a thousand workers and transferring the work to other factories—mostly outside the United States.

We aren't losing manufacturing jobs because we are getting better at this stuff. We are losing them because we aren't competing for them, because around the globe, other countries have stepped up to attract companies to build facilities in their cities, while the United States has wrongfully assumed that its status as the world's only superpower would somehow save the day. An Economic Policy Institute study concluded that as many as 1.78 million jobs lost since 1998 in manufacturing were due to the trade deficit—the direct result of competing against imports from other countries. Any suggestion otherwise is simply not tethered to reality.

Adding Value the Only Way We Can

The followers of the conventional wisdom also don't appreciate that we cannot replace our manufacturing sector with services and expect to add sufficient value to our economy.

While there are, of course, some exceptions, in large part, service sector jobs don't add nearly as much value to the economy as manufacturing sector jobs do. An attorney who represents someone in court will provide a much-needed service to the client, and will be paid handsomely for having done so. Money is transferred from the client to the attorney. The client is happy, as is the attorney. But no monetary value has been added to the overall economy as a result of the transaction. There is no product, no material that is worth more after the transaction than before. Money that was once in the hands of the client has simply been transferred into the hands of the attorney.

Of course, service sector jobs do support economic activity, and I don't mean to suggest for a moment that the United States could prosper without them. Service sector jobs are important—to the millions who hold them and to the hundreds of millions who need those services. A thriving economy must be, by definition, a balanced economy. In addition to manufacturing, the United States requires a healthy service sector, and a strong agricultural sector as well. Much of our GDP depends on America's continuing role as the bread basket of the world, exporting food we grow on our farms. I am not suggesting that manufacturing alone is sufficient for growth. Rather, I mean that it is a prerequisite for growth. Compared to other sectors manufacturing adds considerable value, directly and indirectly, to the economy.

In manufacturing, raw materials and resources are given value when they are turned into high-demand products that can be purchased and sold. Think of an American classic: steel. Steel doesn't begin as a high-value product that can be molded into buildings and bridges and cars and trains. Steel begins as rocks. Lots and lots of rocks. These rocks are mined, and the iron within them is extracted. The iron is then chemically manipulated to become the most ubiquitous metal product in the world. It's that kind of process that adds value—the process of taking something of minimal worth, and transforming it into something that people want to buy.

Silicon becomes a silicon chip. Nickel becomes a key component in the battery of a hybrid car. Petroleum becomes chemicals that are used in the creation of everything from desks and computer monitors to airplanes and musical instruments. At each step in the chain, value gets added to the product, and thus added to the economy as a whole. That's how production creates value for the national economy; and it's why a pure service economy alone could never succeed.

Dow's business—chemicals—illustrates this. Consider just one of our many value chains. We take oil and gas, add heat and pressure and a variety of different chemicals and chemical processes to it. The end result is a chemical called propylene. Now even if I went out on the street in Midland, Michigan, where Dow is based, I guarantee you that most people wouldn't know what propylene was. But propylene is, in fact, one of the most ubiquitous chemicals in the world, and offers extraordinary value chiefly because it has so many uses. It's a key component in products as varied as PVC pipes, diapers, chairs, and clothing. In the United States, we are, by far its largest consumer.

But at Dow, this is just the first stage; we're not done with the chemistry just yet. We can then take that propylene and oxidize it to make propylene oxide. We're one of three companies in the world that does that. Just to give you a sense of what propylene oxide could do, if you had a little flask of it with you and lit it, it would make a crater 10 miles wide. It's incredibly dangerous, but it's also a technology-rich, extraordinarily useful product; we use it in the production of a broad variety of plastics.

Now just from that process, we've created value. For every pound of propylene and propylene oxide, we've added 10 jobs with an average salary of $150,000 a year. And we're still not finished yet. Now we go to the next step—something called a polyether polyol. This is the stuff that makes the foam in a chair or a car dashboard. It's used in appliances, airplanes, furniture, bedding, walls, and insulation. That adds even more value—roughly 10 more jobs per pound.

And still, we aren't finished. We do some more chemistry magic and create an adhesive called urethane. 3M is one of our biggest customers for this product. Now we're talking about having taken that initial oil and gas and created a consumer good—the adhesive, the sealant, the tape, the tube. The chain we've created is incredibly value-rich. For every dollar we spend on oil and gas on the front end, we've created $20 of value on the other side. That entire chain, at world scale, would create roughly 200,000 jobs for the industry.

As a general rule of thumb, every job created inside a chemical plant also creates five jobs outside of it. There is simply no service sector equivalent to this kind of job and wealth creation.

And even among those positions outside of manufacturing that do add value to the economy—jobs like engineers and designers—the bulk of the value they add to the economy doesn't result from the idea itself, but from the mass production of it. The blueprint for a new product is only worth something because it will result in the building of that product. And when engineers and designers innovate those products at home, but see them manufactured abroad, the substantial portion of the value-add from their efforts accrues offshore.

Trying to Survive on Ideas Alone

Still, even if it turned out that the believers in the conventional wisdom were right, even if it were the case that the United States could thrive as long as we kept innovating new products, as long as we were the world leader in research and development, the country's economy would still stagnate without a manufacturing sector. That's because you cannot separate innovation from manufacturing. Where manufacturing goes, innovation inevitably follows.

That might sound backward at first glance. Common sense would tell you that manufacturing would follow innovation. After all, a product doesn't get manufactured until it first gets thought of, sketched out, turned into a prototype, and tested. But when companies are deciding where to build their R&D facilities, more and more, it makes less sense to build those facilities far away from the manufacturing plants themselves. An engineer who develops a new prototype is better served walking across the street to get it made than she is sending it across the ocean. And what we are seeing now, as manufacturing continues to move offshore, is that the engineers and designers are moving offshore with it.

As Alan Brown of *Mechanical Engineering* explained, "These plants needed engineers to run the manufacturing operations, adapt global platforms to local conditions, specify parts, and certify local manufacturers. Once companies established an offshore presence and the procedures to move work back and forth, it became easier to do more product support, additional upgrades, complex reengineering, and even next-generation design *where* they made their products." (Emphasis added.) Companies know that putting their innovators next to their producers is usually the more sensible choice.

Brown cites a study performed by Kenneth Kraemer, the former research director at the University of California, Irvine, which found that "the more a company outsourced manufacturing, the more it outsourced the design and development of its products. In other words, manufacturing pulled engineering to the plant."

Take Mark Pinto, for example. When it comes to high tech research and development, Mark Pinto is at the top of his game. After getting his Ph.D. in electrical engineering from Stanford University, Mark began his career at Bell Labs, where he was responsible for leading silicon-related R&D programs. After Bell Labs he spent some time working at Agere Systems, an integrated circuit company, before ending up as the chief technology officer of a company called Applied Materials.

In the years before Applied Materials was founded, there was essentially no market at all for transistors and semiconductors. According to Applied Materials's CEO Mike Splinter, "The first integrated circuits contained just a few transistors and the production costs were about $1,000 per circuit in 1960 dollars." Given the price, the technology was only used by the federal government, which purchased almost every available circuit and used them for defense technology and the Apollo space program. As Splinter notes, "The result of this early market creation by these government programs was that the cost per circuit dropped an astonishing 98 percent to $25 by 1963." That is how the semiconductor industry was born. And it's how Applied Materials, which was founded in 1967, became a true Silicon Valley icon.

Applied Materials makes the equipment that makes semiconductors. Advances in research and development on their end have been translated into huge cost savings for their customers. Imagine this for an amazing statistic: Thanks to Applied Materials' technology advances, a single transistor is 20 million times cheaper today than it was 40 years ago. To put that in perspective, as Splinter describes, had today's $200 MP3 player been produced in 1975, it would have cost $3 billion.

And it's not just the micro level they are working on. Applied Materials has also become the world's leading supplier of solar panel equipment, thanks to a process called SunFab, which Mark helped introduce.

But for all their success in the Silicon Valley, Applied Materials has thrived in large part by shifting their manufacturing processes abroad. And by shifting their R&D facilities with it.

According to *The New York Times*, Mark is the first chief technology officer of a major U.S. technology company to move to China. "In addition to moving Mr. Pinto and his family to Beijing in January," the *Times* explains, "Applied Materials, whose headquarters are in Santa Clara, Calif., has just built its newest and largest research labs [in Beijing, as well]."

Mark Pinto may have been the first chief technology officer of a U.S. company to relocate abroad, but he surely won't be the last. Plenty of other companies are also investing substantially in research and development outside of the United States. According to the Economic Policy Institute, General Motors has built major research facilities in India and China. As of 2007, Pfizer had 44 new drugs under clinical trial that were developed in India. Microsoft operates an R&D facility in India with more than 1,500 employees. Intel established an R&D center in India back in 1999. Intel also recently completed an R&D plant in China, which CEO Paul Otellini described as being part of a transformation from "manufactured in China" to "innovated in China."

In fact, of the ten U.S. companies that spend the most on R&D, eight of them have R&D facilities in China or India. At Dow, we too have built research and development facilities closer to our production outside of the United States. Just like the other chief executives of these major technology corporations, I understand that I cannot afford to separate innovation from manufacturing. That ignores the complex nature of innovation.

Research and development works best when it can be linked to production. Engineers on the factory floor are more likely to notice potential areas where a product can be improved while participating in the production process. But companies like Dow are not just building R&D centers overseas so that we can narrow the proximity between our builders and our designers. We are also doing it because the incentives we are offered to do so are substantial.

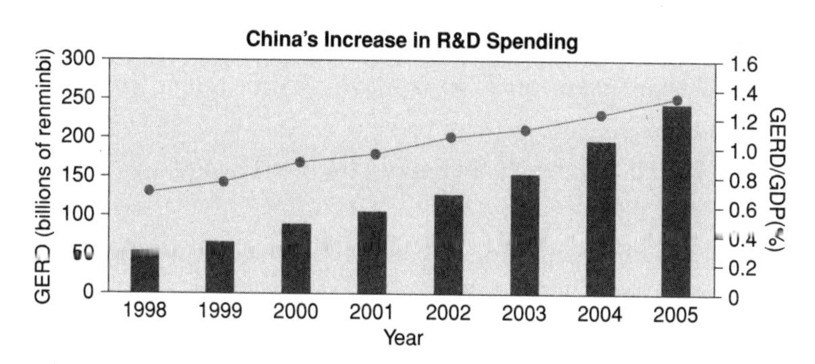

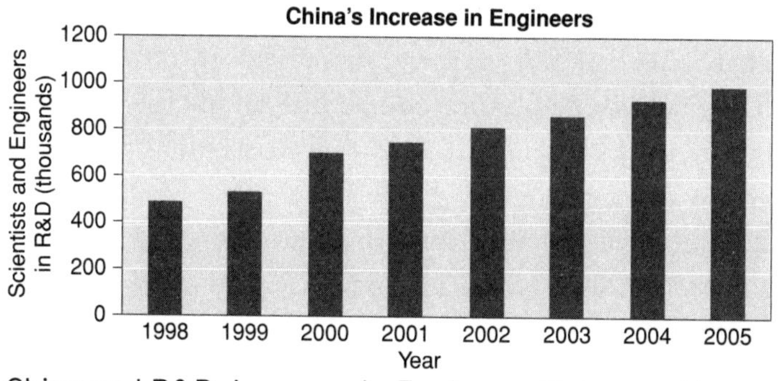

China and R&D: Increase in Engineers Year over Year

SOURCE: Reprinted with permission from "China's 15-year science and technology plan." Copyright © 2006, American Institute of Physics.

The debate that is happening in the United States about whether we should embrace our manufacturing sector or just focus on innovation is truly a foreign concept to a lot of our major competitors. Other countries don't want to do just one or the other; they want to do both. China isn't deciding between growing its manufacturing base or building its innovation capacity. Neither is India or Brazil. These countries don't just offer incentives to attract manufacturing facilities; they offer them for R&D centers too.

In addition to offering lower costs for labor and capital, developing countries offer government subsidies, tax breaks, and other incentives. China sometimes even requires companies to build R&D facilities in order to have access to the Chinese market. And it works. According to the Chinese Ministry of Commerce, in 1999 there were only 30 research institutions in the country that had been established by foreign multi-national corporations. By 2005, that number had climbed to 700. By 2008, it was nearly 1,200.

But our competitors aren't just looking to attract foreign investment in R&D. They are also working hard to build their indigenous capacity for innovation. Over the past decade, the Chinese government has been dramatically ramping up its R&D spending. Though the United States continues to spend more than any other country on R&D—maintaining our one competitive advantage—China is working hard to catch up. By 2002, they were spending more than Germany. By 2006, they were spending more than Japan. And the Great Recession will just make those differences more pronounced. During 2008 and 2009, China and India had a combined 7.6 percent growth in GDP. The 38 other R&D-spending countries saw a decline of 3.6 percent over the same period of time.

Just in the city of Xi'an alone, the Chinese are operating 10 major research facilities as big as Bell Labs. And they are working hard to attract their best scientists and engineers back from the United States.

In 2006, the Chinese government began a 15-year plan aimed at transforming the country into an "innovation-oriented society" by the year 2020. To achieve its aim, the plan calls for China to increase its R&D investment by more than 85 percent by 2020 and to limit its dependence on foreign imported technology to 30 percent. China is also investing in a program called *Qianren Jihua*, which, according to *Science Magazine*, "aims to recruit up to 2,000 top-notch scientists, entrepreneurs, and financial experts from abroad over the next 5 to 10 years." Neither China nor other countries aiming to compete in the global economy are willing to cede innovation *or* manufacturing. It's time for the United States to wake up to that reality, to look around to other countries and see, finally, that if we give up on manufacturing, we will be giving up on innovation. We cannot possibly afford to lose both.

Where Manufacturing Goes, the Ideas Follow

We know that the service sector will never add enough value to our economy to make up for the loss of manufacturing. We know that even if we could somehow, some way, support enough jobs at home for engineers and developers, the vast majority of our population lacks the technical skills to become either. We have plenty of skilled workers that can add real value to our economy. But we can't expect the workers on the factory floor to suddenly become the engineers in the R&D facility next door. It's simply not possible.

When an economy moves to a pure service sector, there is no technology that moves from the service sector back to the manufacturing sector, since there isn't much of a manufacturing sector at all. So if industry moves offshore, a pure service economy has no R&D to speak of; it lacks the infrastructure to design new products, and the ability to bring those products to market. Without R&D it has no intellectual property generation. When that happens, the country finds itself relying entirely on someone else's intellectual property. At that point, America's reign as economic superpower will essentially be over; we will be without innovative new ideas, and without the means to turn them into valuable products.

The frightening truth is that we are closer to that moment than it might seem. The National Science Foundation released a report in 2010 that found that only 9 percent of U.S. companies participated in product innovation between 2006 and 2008. As the report points out, only 7 percent of the companies without R&D reported a product innovation over those years. With innovation on that severe of a decline, the notion that we can survive on our ideas alone clearly deserves to be reconsidered.

The evidence is overwhelming: the manufacturing sector is in severe decline and things are just getting worse. Giving up on the sector, aiming to replace manufacturing jobs with service sector jobs will never create the kind of wealth we need to grow our economy. And it will mean losing our innovation edge too, as the R&D centers where the next big ideas emerge follow the shuttered plants and factories offshore.

If we do nothing, we will be left with nothing.

As economist Michael Spence wrote in *The Financial Times*, "There is little doubt that America's social contract is starting to break. It had on one side an open, flexible economy, and on the other the promise of employment and rising incomes for the motivated and diligent. It is the second part that is unraveling." We are, indeed, emerging from the recession with an economy that cannot live up to that deeply American promise: play by the rules, work hard, and you will have a job and the means to support a family.

As Intel co-founder Andy Grove has written, "You could say, as many do, that shipping jobs overseas is no big deal because the high-value work—and much of the profits—remain in the U.S. . . . But what kind of a society are we going to have if it consists of highly paid people doing high-value-added work—and masses of unemployed?" That might work for small nations and small economies. But the United States has an economy that is simply too large in scale, and too complex, to provide sustainable employment without manufacturing at its core.

There isn't a shortcut to solving this problem. We cannot fix our manufacturing crisis by deciding it isn't actually a problem. Nor can we expect to become more competitive in the global economy if we assume that everything will work itself out in our favor, whether we act or not. We cannot decouple manufacturing from innovation; we cannot separate what can't be separated. And so in order to succeed in both, we must turn our attention toward dealing with the mounting crisis we have on our hands.

To bring manufacturing back to our shores, we have to compete for it. We have to fight for it. We have to do things that the United States is currently ignoring at its peril.

Chapter 3

Fighting Offshoring

If you look inside any electronic device in your home or office—a computer, cell phone, iPad, radio—you'll see an array of microprocessors that communicate with each other and tell the device what to do and how to do it. The key component of each chip—the brain—is a semiconductor, a well-known innovation that's driven revolutionary changes over the past two decades.

When it comes to semiconductors, there is no margin for error. For an electronic device that uses microchips to work correctly, its semiconductors must be flawless. Even the tiniest imperfection—at the molecular level—will slow down or shut down the device.

So if you're in the business of manufacturing semiconductors, or anything that uses them, you need to test those semiconductors for efficiency, consistency, and long-term durability. In the 1990s, a California-based company called FormFactor came up with a great way to do just that.

FormFactor was founded by a Ukrainian immigrant named Igor Khandros. As an engineer at IBM, Khandros had developed a device that did an excellent job of testing the quality of semiconductor materials. He left IBM, found some investors, and started his own company. Today, 15 years later, FormFactor brings in about half a billion dollars in sales annually.

As Clyde Prestowitz of the Economic Strategy Institute notes in *Manufacturing a Better Future for America*, "When American leaders talk about the future of the U.S. economy, [FormFactor] is the kind of company they describe: entrepreneurial, high-tech, high-value added and globally oriented." Indeed, FormFactor's journey is a quintessentially American story—a gifted immigrant uses his brainpower and entrepreneurial spirit to build a small empire. But that's not where the story ends. Instead, FormFactor took another journey—one that is also becoming quintessentially American—from California to Singapore.

A few years ago, according to Prestowitz, a Korean company—Phicom—sprouted up and began making a product similar to FormFactor's. By similar, I mean identical. It was a clone. FormFactor sued Phicom in Korea for patent violations. The Korean court, however, ruled in Phicom's favor.

FormFactor now faced a very serious problem. A major competitor had essentially stolen its product and was selling it for much less than FormFactor possibly could. For FormFactor to compete, even to survive, it would have to invest substantial sums in R&D, hoping to leapfrog the Koreans by developing the next generation in semiconductor testing technology. But FormFactor didn't have that kind of cash on hand.

I can tell you, this is the kind of crisis that haunts CEOs. It raises questions—you can call them existential questions—that lack clear or easy answers. What to do? What options are available? Can my company survive?

As Prestowitz recounts, FormFactor was grappling with these problems when it got a call from the Economic Development Board of Singapore. The board, empowered by Singapore's government to attract foreign investment, came to Khandros with an offer: Move your company to Singapore and we will provide you more incentives than you can imagine. Tax holidays, free property on which to build a plant and headquarters, substantial capital grants for equipment and materials—the incentives were powerful indeed. And the savings, as FormFactor could see, would allow them to invest in R&D without hurting the company's profitability.

Khandros didn't want to accept the deal. He had made his fortune in the United States and felt a genuine sense of loyalty to the country that had allowed him, an immigrant, to live the American dream. As an Australian who has found similar success in America, I understand that feeling well.

Khandros tried, in earnest, to stay in the United States. He turned to the federal government and asked if it was possible to receive comparable incentives here. He was met with indifference. As Prestowitz notes, "The officials with whom he met essentially brushed him off, saying that the U.S. government could not offer FormFactor any assistance in allaying the costs of expanded research needs or protecting its intellectual property rights."

And so, feeling that he had no choice in the matter, Khandros reluctantly did what so many CEOs of American manufacturing companies have had to do: he accepted the reality that the United

States could not compete for his business. He closed his U.S. operation and re-opened it in Singapore. Today, when U.S. companies buy a FormFactor device, it adds to the trade deficit.

Unfortunately, FormFactor is not alone. Not by a long shot.

Thousands of other companies, large and small, old and new, are finding the business environment in the United States unsupportive at best, suffocating at worst. They face financial realities that, in time, force many of them out of the country, even when they wish to remain.

Intel, the microchip giant, has moved many of its operations offshore for similar reasons. In March 2010, *New York Times* columnist Tom Friedman interviewed Intel's CEO, Paul Otellini, about the issue. "A new semiconductor factory at world scale built from scratch is about $4.5 billion—in the United States," Otellini explained. "If I build that factory in almost any other country in the world, where they have significant incentive programs, I could save $1 billion." Indeed, according to the Semiconductor Industry Association, building a chip plant in the United States adds, on average, a billion dollars in costs over its life.

Faced with numbers like that, is it any wonder that Intel built its most recent factory—for, as it turned out, $2.5 billion—in China?

Should I Stay or Should I Go?

Some companies are still waging a valiant effort to stay in the United States, but it can be a constant struggle. Take Bridgelux, for example, a start-up company with a truly amazing new product—one that may well create an entirely new market. The company builds small, light-emitting chips that are only one hundredth the size of a typical light bulb, but emit the same amount of light using 80 percent less electricity. That means they can last 19 years.

In 2009, as Bridgelux was searching for a new site to build its chip facility, the company considered the costs of building in the United States—and elsewhere.

"Look at Singapore," said Bridgelux CEO Bill Watkins. "They say we'll pay for half your manufacturing plant . . . Why can't we do that here in the U.S.?"

"Other countries actually pay you to create jobs," noted Watkins, understandably frustrated. "The rest of the world is chewing us up alive."

Ultimately Bridgelux was able to take over a plant in Livermore, California. According to Watkins, the plant will be the first new semiconductor operation in the Bay Area in a quarter century. But the company wouldn't have been able to build it without having raised an additional $50 million in capital—something no company, not even the great ones with great products, can count on in a pinch. And even then, the fate of the Livermore plant is uncertain. Watkins told *BusinessWeek* that all he would need for the plant to get off the ground were a few big customers (ideally state or local governments) to commit to buying enough LED lights to allow the operation to be scaled up. "So far," *BusinessWeek* noted in July 2010, "Watkins has made 'zero progress' on getting help for his Made in the U.S.A. plan."

As a result, Bridgelux will move forward with projects in Asia and elsewhere, while Watkins waits for an American commitment that may never come. Bridgelux has Asian government contracts to retrofit streetlights and office buildings. One Asian country even offered to pay 80 percent of Bridgelux's workers' salaries for a decade. Sweetening the deal even further, the government offered tax breaks, low interest loans, and free land.

Watkins hasn't given up on Livermore. He's hoping to invest another $150 million. "I'm going to spend the next year trying to find someone that is willing to pay to have something made in the U.S.," Watkins says, noting that when he mentions such hopes to foreign governments, "they laugh. Everyone knows the costs don't work."

Another company facing a similar challenge is a solar energy start-up called MiaSolé. The company manufactures ultra-thin solar cells that can generate electricity much more efficiently than standard cells. Dow is working on a similar product too.

Demand, unsurprisingly, is rising, but MiaSolé's plant is not equipped for high-volume production. When the company's CEO, Joseph Laia, considered where to build a facility that could meet his production demand, he faced a tough choice. He told *BusinessWeek* he'd prefer to stay in the United States, but that "10-year tax holidays in Asia are really hard for a board to get around." Laia added, "The political guys in Washington don't have their minds around the fact that the climate for manufacturing here is really hostile."

Ultimately, MiaSolé, like Bridgelux, managed to remain in the United States, and built its high-volume plant in Georgia. Its ability to do that was entirely dependent on a one-time tax credit that had been part of President Obama's recovery package—the kind of incentive you get from the federal government only when the economy is in the doldrums. The Obama policy made MiaSolé eligible for more than $90 million in tax credits, enough to offset the kind of incentives offered by other nations in both good times and bad.

This piece of good news—for MiaSolé and America—offers little encouragement to other U.S. manufacturers. The tax credit was capped; three times as many companies have requested it as there is money for it. MiaSolé was fortunate. Others may be forced to try their luck overseas.

It Isn't What You Think

Why are U.S. companies forced to go abroad with such frequency? Why is it so much more expensive to do business at home than abroad?

The answer you hear most frequently from pundits and politicians alike is "labor costs." There's a widespread view that the deck is stacked in favor of developing countries because their people are willing to work for pennies. How can a U.S. manufacturer compete, the argument goes, if our workers make $20 per hour and Chinese workers make $2?

It is certainly true that wages play some role—sometimes a significant one—in the decision to build a plant overseas. When Maytag moved its operations from Iowa to Mexico, cheap labor was the main reason. The same consideration led Levi Strauss & Co., the iconic American jeans maker, to relocate from the United States to China and Mexico.

But it's a mistake to assume that, in general, manufacturing jobs are leaving the United States chiefly because of cheap labor elsewhere. For one thing, low wages in developing countries are often offset by higher productivity in the United States. Let me explain. If it takes 10 workers in a developing country 10 hours each to make a product, at a wage of $3 per hour, then your labor cost for each product is $300. If you build that product in the United States instead, it might take three highly skilled workers only three hours each. Even if the American workers earn $30/hour—10 times the rate in the developing country—your total labor cost is $270. That's 10 percent less than you'd pay overseas.

America has the most productive factories in the world. As Richard McCormack explained in *The American Prospect*, it takes the Chinese steel industry six times as many hours as its U.S. counterpart to produce a single ton of steel. And the Chinese emit three times as much carbon to get the job done. The United States will never be able to compete dollar for dollar with wage rates in developing countries. Nor should we. But our productivity substantially narrows the gap.

Besides, if cheap labor were the primary reason to leave the United States, every American company would be doing its manufacturing in Honduras and Burundi and Sierra Leone. If cheap labor were the driving force behind where to locate a manufacturing plant, even China would be in trouble. The average wage rate in China might be lower than in the United States, but it is still higher than in dozens of other countries.

As New York University business professor Ralph Gomory wrote on the Huffington Post,

> Cheap labor doesn't explain the fact that Japan and Germany, both high-wage countries, are successful in the automobile industry. Nor does it explain how semiconductors, a model of high investment, low-labor content industry, are mainly made in Asia. The premise that the inescapable burden of competing against low wages means failure is simply not correct.

Indeed, the manufacturing success of a country like Germany, which has wage rates that are competitive with the United States', should be proof enough that cheap labor cannot explain why U.S. manufacturing is struggling. But there's no denying the perverse appeal of the conventional wisdom: if starvation wages are the issue, if that's what it takes to compete, then we can let ourselves—American business leaders, policymakers, all of us—off the hook.

But the facts, as I've argued, don't support this passive approach. Countries with higher wage rates can compete—and prevail—against countries with substantially lower wage rates, and that's good news for America. The bad news for America is that other nations are working hard to gain other kinds of competitive advantages—and here, again, the U.S. response has been a shrug of the shoulders, a grim acceptance of what seems to be our fate. Often that's paired with a complaint that other countries aren't playing fair.

Sometimes that's true, of course. Nations don't always follow the rules, from trade and intellectual property laws to currency devaluation. But the real issue, the real source of many Americans' frustration, is that the rules are changing—and that certain countries are getting very good at working those rules to their own advantage. China, India, Brazil, and others have spent years watching nations like the United States—learning from our success, refining and outmatching our strategies for attracting private investment and creating jobs, and looking carefully for opportunities we might be missing. They seek to differentiate themselves from their competitors. They are agile. Aggressive. Creative. Relentless. They offer generous incentives and the promise—or at least the strong prospect—of a better return on investment.

In other words, globalization has made countries behave a lot like companies.

They take into consideration what my biggest concerns are as a CEO, where my biggest risks exist, and they work to minimize them. When I go into a boardroom to discuss the site location of a new facility, there are a number of variables to consider: the costs of inputs, the costs of building the asset, the market I can sell to. Do I have the human capabilities, the scientific and research capacity? Can I invest in R&D successfully? Can I scale up? The answers to these questions matter. They are the difference between losing our shirt and making a lot of money for our shareholders. If you're putting billions of dollars of shareholder money at risk, your thinking process is always the same. "I need to minimize the risk for the shareholder. I need to get the best inputs from the best people, believe them, believe it myself, then convince my board." These are the issues we care about.

Enter stage left a government that says to you, "I will give you 20 years of discounted inputs fixed, and the discount is 90 percent off market." Or, "I will give you a tax break and a low-interest loan, and half the capital up front for building the site." Or, "We'll give you a fixed price on your inputs, a contract price to control for volatility." They just eliminated a very big risk for me compared to an alternative investment in the United States. They are removing my uncertainty. And they aren't doing it strictly for my benefit. They know it will mean jobs, both direct and indirect, in their country. They can justify to their financial system the lower prices because they know that getting me to build in their country at less than world cost will create revenues up and down the chain that will more than cover that discount.

When I go to Germany, Angela Merkel asks me what Germany can do to attract more Dow investment. She knows that our further entry into her country will create entire value chains of sustainable jobs. She knows what that could be worth to her economy. But in the United States, these are not the kinds of questions you ever get asked by national political leaders.

The American government's unwillingness to aggressively compete in the global economy as others have doesn't just result in companies moving offshore. It results in entire industries disappearing. Until the year 2000, for example, the U.S. chemical industry was the country's second-largest export sector. But volatility in our raw materials changed all that. Where other countries gave chemical companies stable contract prices on their feedstocks—as a way of bringing a measure of certainty to the equation—the United States said, once again, just let the markets rule. And the markets ruled. Half a million chemical manufacturing jobs moved offshore.

At a time when presidents and prime ministers are sounding a lot like CEOs—when they ask themselves, "What can we do to win this piece of business? How can we sweeten the pot? What will it take to close the deal?"—then we know the old answers no longer suffice. The United States has long taken it for granted that other nations envy and emulate us. They still do—for very good reasons. But it's time for us to acknowledge that some of the lessons they've learned from us are lessons we're at risk of forgetting ourselves.

Taxing Problems

On the morning of October 22, 1986, President Ronald Reagan stepped out onto the South Lawn of the White House. Assembled there were senior administration officials, members of Congress from both parties, and a group of CEOs. They had gathered to watch the president sign the Tax Reform Act of 1986, a bill that substantially lowered tax rates for families and businesses alike.

"In a moment I'll be sitting at that desk, taking up a pen, and signing the most sweeping overhaul of our tax code in our nation's history," Reagan began. Among the bill's provisions was a reduction in the corporate tax rate from 46 percent to 34 percent. The logic was that companies had a choice of where to operate, and if the United States could offer a lower corporate tax rate than other nations, it would make America a more attractive place to do business, create jobs, and generate economic growth.

"When I sign this bill into law," Reagan continued, "America will have the lowest marginal tax rates and the most modern tax code among major industrialized nations, one that encourages risk-taking, innovation, and that old American spirit of enterprise. We'll be refueling the American growth economy with the kind of incentives that helped create record new businesses and nearly 11.7 million jobs in just 46 months."

The efforts of President Reagan and a bipartisan coalition in Congress paid off. Indeed, by 1988, when the Tax Reform Act had taken full effect, the U.S. corporate tax was about 12 percent below the average rate among G-7 nations. And with that came new business and growth, just as anticipated. American manufacturers built new plants and created new jobs.

But the success of 1986 didn't last for long. The United States wasn't alone in recognizing that low tax rates would attract corporate investment. Other countries could see the power of that incentive. What followed was an arms race of sorts, with countries competing aggressively to become the most attractive spot for corporate investment. After the Reagan tax cut, every single developed nation lowered its corporate tax rate. And plenty of developing nations—including China, Singapore, Mexico—did too.

For all the talk of cheap labor luring jobs offshore, you'd never guess that companies like HP, Microsoft, Dell, and Intel would choose to build manufacturing facilities in Ireland, where the average wage rate in 2007 was about $19 per hour. But that's exactly what happened—because in the mid-1990s, Ireland lowered its corporate tax rate to 12.5 percent, by far the lowest in the world. When Microsoft opens up a facility in Ireland, it pays just over half the taxes it would elsewhere in the EU, and almost 70 percent less than in the United States.

According to the Cato Institute, the average corporate tax rate in Europe has dropped 14 points since 1996, to a level of 24 percent. But during that same period, the United States actually raised its rate to 35 percent, which is now well above the G-8 average. When state taxes are included, a corporation operating in the United States faces an average statutory rate of 39.1 percent. Today, even when tax credits—which somewhat offset the taxes—are included, the United States has the second highest corporate tax rate in the world. Only Japan's is higher.

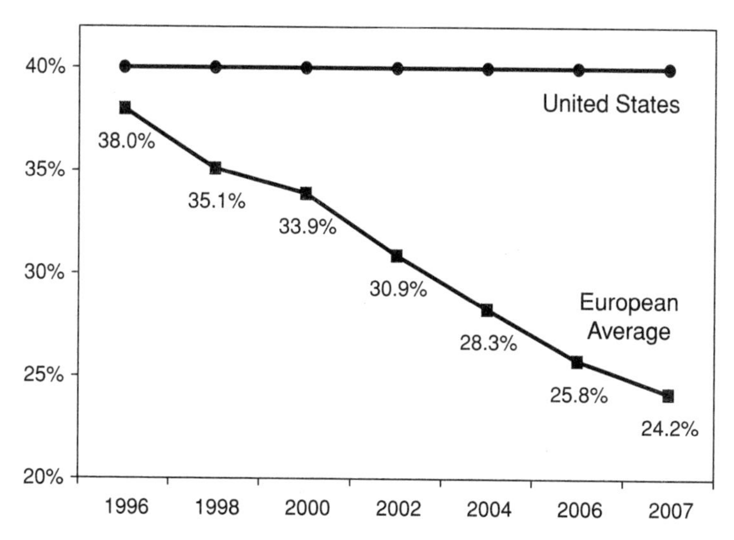

Corporate Tax Rates

SOURCE: CATO Institute: Corporate Taxes: America Is Falling Behind.

DATA SOURCE: KPMG. Data include both national and subnational taxes.

The United States touched off this race—then dropped out. What happened? In part, the politics of tax cuts changed over time. "Reduce corporate tax rates!" is not a great rallying cry, particularly in tough times, and especially when the cry is issued by, well, people like me. Politically speaking, it's more palatable—and rewarding—to reduce personal income tax rates. And while politicians of both parties sing the virtues of cutting taxes, few acknowledge that high corporate taxes hurt the typical American family by putting their jobs at risk.

When a multinational corporation is deciding where to locate a new plant, one of the things that any competent CEO has to consider is the cost of doing business in America. Tax rates are a big part of that calculus. When U.S. rates are uncompetitive, plants get shut down in Ohio and reopened in Brazil. Start-ups build their future on Chinese soil. So the people hurt most by the high corporate taxes aren't the corporations themselves, but the American people—those who work in the plant, or somewhere else along the supply chain, and who will suffer when the manufacturing sector stagnates.

One argument against lowering corporate taxes is that it would increase the federal deficit. There's an obvious logic at work here. The federal government collects billions of dollars every year in corporate taxes; if we cut the rate, it stands to reason that tax revenues will drop.

But in reality, cutting the corporate tax rate can actually *increase* tax revenue. A lower tax rate attracts investment, both foreign and domestic. What the government initially loses in revenue is made up for by the new business the lowered rate attracts to the country.

A study by the World Bank explains that "high tax rates do not always lead to high tax revenues. Between 1982 and 1999 the average corporate income tax worldwide fell from 46 to 33 percent, while corporate income tax collections rose from 2.1 percent to 2.4 percent of national income." In its own report on the issue, the European Union explained that it is "quite striking that the decline in corporate income tax rates has not resulted, so far, in marked reductions in tax revenue, [with] both the Euro area and the EU-25 average actually increasing slightly from the 1995 level."

Of all the things in which to be a global leader, corporate taxes shouldn't be one of them. And this, unfortunately, is not the only area where the United States was once doing the right thing, and has since sacrificed its competitive advantage to other nations.

Funding the Future

The United States, to its credit, was one of the first countries in the world to make innovation a top priority. In 1981, an R&D tax credit was introduced, which companies can use to reduce the cost of innovating new products and experimenting with new ideas. The R&D tax credit has been a strong incentive and has been used by businesses that have created millions of jobs and a number of essential new industries.

But here, again, other countries were paying attention. Seeing the success of America's R&D tax credit, they began not only to replicate it, but to outdo it. According to Robert Atkinson, an economist and the president of the Information Technology and Innovation Foundation, "in the last decade every country that has an R&D tax incentive has increased the generosity of those incentives." In fact, he points out, by 2004 the United States had fallen to seventeenth in R&D tax generosity among Organisation for Economic Co-operation and Development (OECD) nations.

When Congress expanded the credit in 2005, the United States did not improve its position—it merely halted the slide.

This means there are now 16 countries that offer substantially better innovation incentives than the United States. In my home country of Australia, for example, a business can deduct 125 percent of its R&D expenses. (That means for every $1 million spent on R&D, a business can deduct $1.25 million. That's substantial.) In France, a company can qualify for a 30 percent credit, which is five times the credit being realized by companies in the United States. According to Ernst & Young, "The average company that claims the U.S. R&D credit only realizes a credit rate of 6 percent" primarily because "the United States requires that the deduction for R&D expenses be reduced by the amount of any R&D credit, which drives the effective rate even lower."

The problem is compounded by a sense of uncertainty. America's R&D tax credit has always been temporary—designed to expire unless Congress renews it. Congress has failed to renew it eight times since 1981, including 2010. Each year, businesses have to face the distinct possibility that the R&D tax credit will be suspended, drastically cut, or discontinued. This makes it awfully hard to plan for the future.

When Dow considers expanding its operations, we face, as you'd expect, a major decision. It doesn't come cheap. We are talking many billions of dollars. Clearly, you can't make a decision on this scale and then change your mind about it the following year. We look at projects like this on a 30-year time horizon, typically, and base our investments on what we expect the returns to be over many years. But the temporary nature of the R&D tax credit means it's difficult to estimate our costs and returns over the long term. If Dow's board of directors were to ask me whether Congress will renew the R&D tax credit every year for the next 30 years, the best and most honest answer I can give is, "I hope so." But the stakes are too high for wishful thinking.

Regulating Our Way into a Muddle

The trouble goes well beyond taxes and tax credits. Making things in America means complying with a wide and shifting array of federal and state regulations.

Now, there are quite a few CEOs out there who argue that regulations are nothing but a burden on companies and consumers—that the market is a sufficient check on corporate behavior. I'm not one of those CEOs. Yes, I'm a strong believer in the power of capitalism to improve people's lives, and I put great faith in markets to determine supply and demand, and to separate winners from losers. But knowing the value of markets requires understanding their limitations. Sometimes, though people in business are loath to admit it, they need the government's help in functioning better. Sometimes markets fail to place monetary value on things that we, as a society, value in a very real way—clean air and water, product safety, workplace safety, fairness, justice.

When regulations are too lax, bad things result. When the country's biggest banks were able to take unjustifiable risks with shoddy financial products, it was largely because of regulatory failures. And when the financial sector collapsed and threatened to drag the United States into a second Great Depression, no one could deny that smarter regulations might have staved off the crisis. More recently, investigations of BP's oil spill disaster revealed that much of the problem was due to the negligence of regulators.

Regulations matter. And one of the ways that the United States stands out in a very positive way from many other nations is its commitment to the health and safety of its people and its workers. Clear, consistent, and wise regulations not only protect the well-being of the population, but improve U.S. industry's ability to function and grow, which benefits every American.

But that's the key point. Regulations are beneficial only when they're clear, consistent, and wise. And in large part, the U.S. regulatory regime is so complex and inconsistent that regulations hinder American manufacturers—without helping anyone in particular.

Here's one of my favorite examples of the inconsistency of regulations. When we at Dow work with oil, as we do every day, we know there are federal regulations we have to follow. I'm sure you'd expect that. But you might not expect that the federal government has three completely different definitions for "oil." The Coast Guard has one definition, the Environmental Protection Agency another, and the Department of Transportation yet another. Which one is the right one in any given case? I'd have to ask Dow's general counsel to know for sure.

This isn't an isolated example. A renewable energy company that believes it might be eligible for a federal rural energy grant, which provides up to 25 percent of rural clean energy project costs, first has to find out if the site location counts as rural. The problem is that the U.S. Census Bureau defines the word differently than the Office of Management and Budget does, which in turn defines it differently than the Department of Agriculture. So which one counts? I couldn't tell you.

This is no way to run a government. In addition to making the regulations themselves less effective, it causes companies to spend unnecessary sums of money on data collection and compliance procedures.

And that's just federal regulations. Many, if not most, of the regulations we are asked to comply with are written in state capitals, not Washington. That creates a patchwork of regulations that force a business to operate differently in California than it would in Michigan, or to make the product it sells in Nebraska somehow different than it is in North Carolina.

I read an article in the *Washington Post* a few years ago about a Seattle-based company called Matter Group, which makes children's games and stuffed animals out of recycled materials. Amy Tucker, the company's president, told the *Post* she was thrilled when Washington State banned the use of dangerous chemicals—including lead—in children's products.

Then Tucker found that the news was not all good for her company—or, in the end, for children. After Washington State changed its laws, other states did, too, in their own variations on the theme—creating a difficult-to-navigate regime that forced toymakers to produce different toys based on different standards for different states. "It puts manufacturers in the position of having 50 different sets of regulations to abide by," Tucker reflected, "and that can become very onerous for a company." The multiplicity of state regulations meant that even some safe toys might disappear from the market: as Cater Keithley, president of the Toy Industry Association, said, "there isn't any one state that has a big enough market to justify the kind of volume production required to bring toys to the market at a reasonable price point."

Toy manufacturers didn't ask to be let off the hook. They didn't ask for deregulation—they asked for smart regulation. "A national product safety standard," said Toys R Us CEO Jerry Storch, "would avoid all this confusion, all of this complexity."

This problem has been especially hard on the auto industry, where the patchwork of standards for greenhouse gas emissions requires engines to be built differently if they are going to be used in vehicles driven in California than if they were made for neighboring Arizona.

The same goes for fuels. The requirements for gasoline vary from state to state, requiring oil companies to have more than a dozen different formulas. That doesn't just affect the companies; it affects the consumer. If there is a problem at a refinery in California, the state can't just turn to a neighbor—say Nevada—to get its gasoline. Nor can the oil company easily have another refinery—which is already at full capacity—start producing California's specialty fuel. The supply goes down, and the price for consumers go up.

Now, I know the private sector has done plenty of campaigning over the years against regulations of many different kinds; but here, again, is an example of companies—in this case automakers—going to Congress to ask for clear, consistent national standards. The state-by-state rules are so time-consuming, confusing, and costly that car companies are actually calling on Washington for more federal regulation.

Often, the problem is unfinished regulations. Congress passes laws that aim to regulate industry, but it's the agencies themselves that actually write the regulations. Yet sometimes, long after Congress has passed the law, the agency in charge doesn't do its job.

I'll give you an example. In 2000, Congress passed and President Clinton signed into law the Transportation Recall Enhancement, Accountability and Documentation Act, or TREAD. The act was designed to set newer, higher standards for tire manufacturers. But the National Highway Traffic Safety Administration, the agency responsible for implementing TREAD, took an extraordinary amount of time to write the actual regulations that manufacturers were supposed to follow.

In 2007, Michael Wischhusen, director of industry standards for Michelin North America Inc., complained about the process and its impact on the industry. "Congress ordered rulemaking for the TREAD Act to be done in two years. It's been seven, and we aren't through the half of it." Wischhusen said that there are people who started working in the tire industry the year the law was passed who will retire from the industry before the rulemaking is finished.

These problems add up. Without fast-paced rule-making and clear, streamlined regulations, the process can easily get bogged down and cause delays for manufacturers at critical moments. Often, manufacturers miss important windows of opportunity while they wait for outmoded approval processes with little, if any, impact on the health and safety of the American people.

This puts the United States at a competitive disadvantage. A World Economic Forum survey of 13,000 business executives worldwide found that there are 52 countries with less burdensome government regulations than the United States. That adds to business uncertainty. Frankly, I'd rather a system with a few bad regulations, then with constantly changing regulations. At least that way I know what to expect, and I can prepare for it.

How we operate within our own borders, what we require of businesses here, often puts us at a competitive disadvantage internationally. But it's also how we interact with other nations that impairs America's economic growth.

Trading Our Way to Prosperity

For many decades, the greatness of American manufacturing—the measure of its power and success—has been the size of its overseas markets. Americans rightly took pride in the fact that you could find Ford automobiles in Europe, or (as I can attest) Maytag appliances in Australia, or that if you were lucky enough to take a trip to Japan, you would see American clothing on store shelves. And no one took greater pride in that than the workers who made those products. They knew they and their companies were making a vital contribution to economic growth here at home and around the world.

Indeed, in the half century that followed World War II, America became the world's best example and advocate of the power of trade, and trade became the greatest force for prosperity, opportunity, and upward mobility on every continent. You see this today in the talks among Pacific nations to develop the next generation of free trade agreements. You see it in the eagerness of Southeast Asian nations, who recently created a free trade zone for their goods, and in the calls one hears increasingly in Africa for "trade, not aid." These countries see trade, more than any other economic activity, as the key to long-term economic prosperity.

The United States, meanwhile, has grown disillusioned with trade. At the very least, we view it with ambivalence. Many Americans have become convinced that trade is not the answer to what ails U.S. manufacturing—some even believe that trade itself is what ails manufacturing.

Washington reflects this same contradiction. On the one hand, President Obama has announced an ambitious goal of doubling exports by 2015. On the other hand, the administration and Congress are doing little to pursue or prioritize free trade agreements. In 2010, the European Union had 11 free trade agreements under negotiation. China had 15 under negotiation. The United States had just one new trade agreement in negotiation, and three others that were stalled, if not dead, in Congress. Other countries are going to bat negotiating access to new markets on behalf of their manufacturers—and at the expense of American operations.

It's time to remind ourselves of what we understood for most of the past half-century: that free trade agreements aren't the enemy of American manufacturing; they are one of its best hopes for long-term success.

It's worth remembering just how important it is for American manufacturers to have places to sell their goods overseas. With 95 percent of the world's consumers living outside of the United States, manufacturers simply must have the access to those markets. Especially when economies like China's, India's, and Brazil's are growing at rates three to four times greater than mature economies like the United States and Europe. The middle class in these countries is growing, too, at a staggering rate. They are going to buy more and more products. Shouldn't many of those products be made in America?

America's massive trade deficit has led us to believe that other nations always get the better of our trade relationships. The equation that defined the previous century—we build it, they buy it—seems to have been turned on its head. But the trade deficit, while real and dangerously large, does not tell the whole story. In fact, the United States actually has a trade *surplus* with the countries with which we have free trade agreements. According to the Manufacturing Institute, in 2008, the United States had a surplus of $21 billion with its free trade partners. But the United States hasn't negotiated free trade agreements with some of our biggest trading partners, and the result couldn't be more stark: in 2008, China, for example, accounted for 62 percent of the total deficit and Japan for 19 percent.

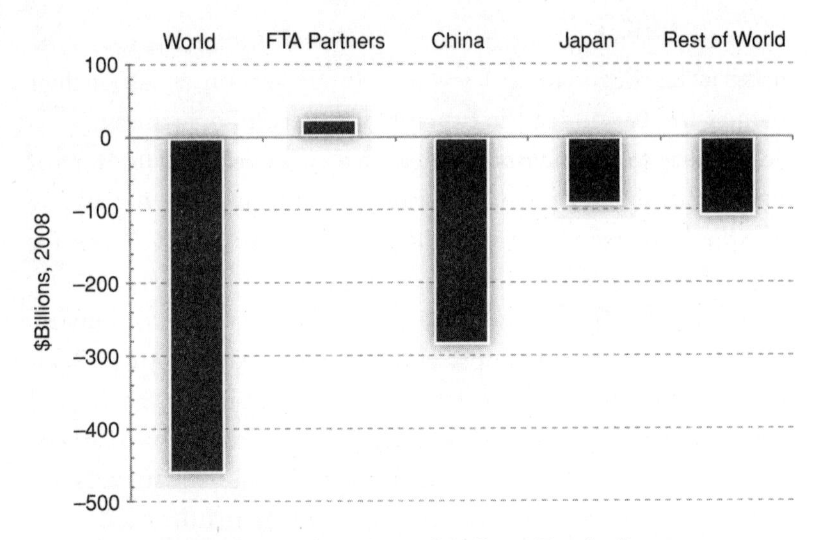

We Have a Trade Surplus with Free Trade Partners
SOURCE: Copyright © 2009 by The Manufacturing Institute. *The Facts about Modern Manufacturing, 8th ed*. 2009.
DATA SOURCE: U.S. Bureau of the Census and NAM press release, May 27, 2009.

Why do we end up at a disadvantage in our relationships with countries like China and Singapore? Partly because their products are cheap, and, relatedly, because our products are expensive. This flows from a number of things, including currency levels and labor rates, among others. But it also has to do with tariffs—the fees that countries add to products that come in from overseas. The United States doesn't believe in trade barriers, and that's a good thing: low or nonexistent tariffs keep trade flowing and keep the cost of imports low, which benefit American consumers. But some other nations—when they're not bound by a trade agreement—see things differently. While the United States imposes, on average, only a 2 percent tariff on manufactured goods that are imported here, our exports to non-free trade agreement partners face an average tariff of more than 10 percent. That's a serious disadvantage for U.S. manufacturers.

Now, I'm not arguing we should increase U.S. tariffs. That would almost certainly spark a trade war and a new wave of global protectionism—not good for anyone. Rather, we want other nations to lower or eliminate their tariffs, reducing barriers rather than building them up. This will take serious engagement and sustained commitment from the U.S. government. But if successful, the effort will be well worth it. There is no doubt that a sound trade policy would facilitate market access and help establish a more level global playing field for manufacturers. It would ensure that our trade partners were committed to critically important multilateral rules that are transparent and fair.

I can tell you what this would mean for my company: approximately 40 percent of Dow's global workforce operates in the United States, supporting American manufacturing, creating products for export, and partnering with global operations for Dow's overall growth. However, almost two-thirds of Dow's sales are outside the United States; in 2009, that amounted to $6.3 billion worth of U.S.-manufactured products being sent abroad. Trade and market access will allow Dow to export even more goods to Dow facilities and customers abroad while broadening its customer base. It would also create more jobs in our American headquarters and more R&D investment. That would be welcome news for our U.S.-based manufacturing.

Dow was rated one of the five most global American companies by *Fortune* magazine. We will no doubt continue to be very active and, we think, successful in overseas markets. But not every U.S. manufacturer has our resources or our history. They are counting on Washington to do what other countries are doing on behalf of their manufacturers—to open new markets to the things they build. These countries have, again, learned a lot from America's success. It's time for America to relearn the lessons it has taught the world. Because if we're not knocking down barriers to trade, we're effectively building them up.

Taken together, these problems add up for American companies. The costs are substantial. They can't be ignored by any entrepreneur. Sometimes they are outweighed by other considerations—among them the very real and powerful desire that many of us have to live, work, create, and contribute in the United States—but often, they are not.

No company, big or small, can possibly ignore numbers like these: a recent analysis by the National Association of Manufacturers found that taxes, compliance, and other structural costs result in, on average, a 17.6 percent disadvantage for U.S. manufacturers when compared to their foreign counterparts. When you're operating on a large scale, when you're talking about a facility that can cost, say, half a billion dollars to build, 17.6 percent equals a penalty of tens of millions of dollars to do it in the United States. You could also look at that as a bonus of tens of millions of dollars to pack up and move offshore. You might decide to forgo it. But the decision's not easy.

I can't overstate the size of this problem—or the lack of understanding that prevails in policy circles. Politicians who are convinced that manufacturing is disappearing simply because of labor costs will never fully appreciate just how many other, more substantial costs are being shouldered by manufacturers.

The good news is that these are problems the United States can solve. Each one, from a dysfunctional tax policy to an often arbitrary

Fighting Offshoring

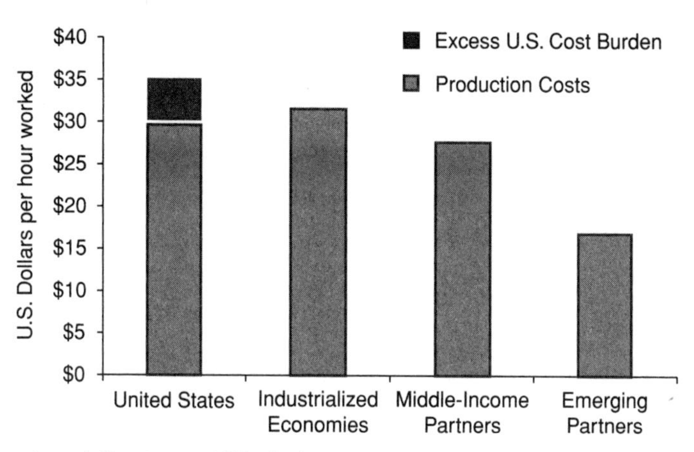

Structural Costs and Their Impact

SOURCE: Copyright © 2009 by The Manufacturing Institute. *The Facts about Modern Manufacturing, 8th ed.* 2009.

DATA SOURCE: The Tide Is Turning, The Manufacturing Institute. National Association of Manufacturers, and the Manufacturers Alliance/MAPI, 2008.

regulatory process, can be addressed and improved. Other nations have started from a position well below the United States in terms of resources, in terms of know-how, and in terms of experience. They have overcome many of those deficiencies through their determination. For the United States, it shouldn't be as difficult as that. America retains incredible natural advantages, and a reservoir of strength and knowledge no other country can match. But we, too, must act with determination.

And we must act as one. The American people need a strong economy—one that can continue to grow, add wealth, add jobs, and improve our quality of living as we march ahead. That isn't a partisan goal. It shouldn't break down along the lines that divide: management versus labor, business versus government, Republican versus Democrat. And if we can agree, as we should be able to, that reviving the manufacturing sector is an essential national project, then we must begin to remove unnecessary burdens from our builders, creators, and dreamers.

Of course, these are not the only obstacles in the path of manufacturing. Over the long term, there are other, broader challenges that America must confront if it is to continue its role as the world's leading economy.

Chapter 4

Energy Drives the World

In March 2010 I went to Houston to address a conference hosted by Cambridge Energy Research Associates, or CERA. The organization, founded by the incomparable Daniel Yergin, advises energy companies and governments on changes in energy markets, new innovations, and, more broadly, the state of energy affairs around the world. The conference, held in the self-proclaimed energy capital of the world, brought together executives and business leaders from some of the biggest oil and gas companies in the world, as well as renewable energy firms. I was there to give the keynote address.

You might wonder why the CEO of The Dow Chemical Company would be asked to speak at an energy conference. After all, Dow doesn't produce oil and gas. We don't operate coal plants. We don't run public utilities. On the contrary: I was there to speak as a consumer of energy. I was there to meet my suppliers.

"Here's a bit of trivia," I began. "The U.S. chemical industry used about $85 billion worth of energy and feedstocks last year. Dow accounted for a lot of that. We use the equivalent of about 850,000 barrels of oil . . . every day. That's roughly the same amount my native country of Australia consumes all by itself."

"So, in one way," I continued, "I suppose I'm here to say . . . you're welcome."

A little humor—to illustrate something important. The manufacturing sector uses an extraordinary amount of energy every year. Energy is one of our greatest costs, and therefore plays a big a role in our sector's overall health and strength. Remarkably, more than 20 percent of all the energy consumed in the United States each year is used by the manufacturing sector. In 2006 manufacturers used more than 4.5 million megawatts—roughly equivalent to the energy used by every home in America combined.

A Big Energy Bill, and Not Just for Power

In Dow's case, not all of the energy used goes to powering our operations. Most companies burn energy, and we do, too, of course. But we also use energy as one of our primary raw materials. We transform it—into more things than you can imagine. Petrochemicals are our raw materials. Just as wood is an essential starting point for a paper manufacturer, oil, natural gas, and other petroleum derivatives are building blocks for chemical manufacturers and, therefore, all the customers we serve.

The molecules we get from petroleum can be used to make the chemicals and plastics that are essential to our lives. They coat the semi-conductors of every electronic device you own. They are in the insulation in your home, the paint on its walls. They are in musical instruments, in tables and chairs, in pharmaceuticals. They are part of packaging, and detergents. They are used in producing everything from candy to candles, from aluminum to asphalt. Louisiana State University professor Ed Overton says it best: "There's nothing that we do on a daily basis that isn't touched by petrochemicals."

The scientists in our labs are constantly working on petrochemical combinations that have never been known before in order to create products that have never been made before. That makes energy issues of particular importance to Dow. In fact, we may care more about energy than anyone who isn't a producer.

Energy costs—both in terms of our power bill and our raw materials—are something we at Dow think about every day. And energy efficiency, as a result, is central to almost all of Dow's manufacturing processes. When you run plants the size of ours, when you power machines at the scale we require, energy efficiency is the fastest, most obvious way to control costs. We have worked incredibly hard to be less energy intensive, and with great success. Since 1994, our company's efficiency measures have saved 1,700 trillion BTUs, which is the equivalent of 13.6 billion gallons of gasoline. We've reduced our carbon emissions by 90 million metric tons, which is the equivalent of taking nearly every American car off the road for a month. And in the process, we've saved more than $9 billion.

This is a critical issue for manufacturing generally. Energy costs can be an unsustainable burden for manufacturers of any kind, for anyone operating a facility with high-tech machinery. In the previous chapter, I talked about structural costs that make it difficult for manufacturers to operate in the United States. Just like those costs, energy prices and volatility often make it more appealing for companies to do business elsewhere, in nations where energy is cheaper and supplies are more stable.

By now we all know that the price of energy isn't the only cost we pay. Over the past decade, Americans have grown increasingly concerned about the consequences of our dependence on foreign oil—from the effects of global climate change to the enrichment of dangerous regimes in troubled parts of the world. We have also become increasingly familiar with the arguments in favor of renewable energy—wind, solar power, biofuels.

I'd like to focus on an argument I find especially compelling: energy as opportunity.

Energy, clearly, is one of the megatrends defining this century. It is only going to grow in importance as global demand increases. Consider this: according to the World Economic Forum, the world's population is expected to grow from 6.6 billion to more than 8 billion over the next 25 years. It's expected to grow another billion in the quarter century after that. That fact, coupled with rising incomes and changing lifestyles in emerging economies, will dramatically increase the need for energy.

The International Energy Agency forecasts that global energy usage will rise 70 percent by 2050, with especially big increases in the industrial sector. In the United States, energy consumption will grow 14 percent by 2035, with fossil fuels providing 78 percent of all energy. Providing this amount of energy, while working to reverse the growth of greenhouse gas emissions, will no doubt be one of the most daunting challenges facing our global society.

I think Thomas Friedman said it best in a September 2009 *New York Times* column: "O.K., so you don't believe global warming is

Industry accounts for close to 1/3 of total end-use energy consumption

Most of the manufacturing sectors' energy comes from fossil fuels with natural gas representing the single most important energy source

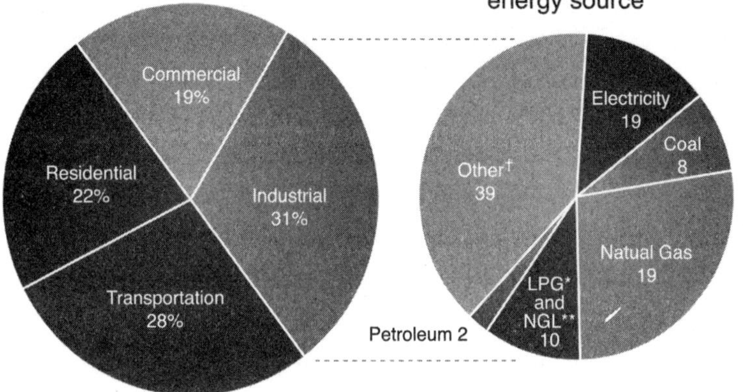

*Liquefied Petroleum Gases

**Natural Gas Liquids

†Other energy includes a mix of energy types (petroleum, natural gas, coal, nuclear power, and renewables). This energy is used to produce heat and power or used as feedstock/raw material inputs. For the petroleum refining industry, the feedstocks and raw material inputs for the production of non-energy products and feedstock consumption at adjoining petrochemical plants are included as "Other" regardless of type of energy.

Energy Use by Sector

SOURCE: Copyright © 2009 by The Manufacturing Institute. *The Facts about Modern Manufacturing, 8th ed.* 2009.

DATA SOURCE: U.S. Bureau of the Census and NAM press release, May 27, 2009.

real. I do, but let's assume it's not. Here," he explained, "is what is indisputable. The world is on track to add another 2.5 billion people by 2050, and many will be aspiring to live American-like, high-energy lifestyles. In such a world, renewable energy—where the variable cost of your fuel, sun or wind, is zero—will be in huge demand."

This is the next frontier of manufacturing. Some of the most exciting work today in the world of manufacturing is being done in the creation of wind turbines and solar cells and batteries. Chemistry is central to many of these solutions, which explains why Dow is so active in this field. In 2010 we broke ground on an 800,000 square foot plant near our headquarters in Midland, Michigan, where Dow chemists, engineers, and technicians will design and build batteries for the next generation of hybrid and electric cars and trucks. When complete, the facility will have the capacity to produce batteries for 60,000 vehicles every year. Also in the Great Lakes Bay region, we've started work on a plant that will manufacture the solar shingles I mentioned earlier.

Dow isn't doing this because we are being charitable. We are a business, after all. We are doing it because it is profitable, significantly so for the companies—and countries—who are first or best able to forge viable, affordable solutions. The advanced battery market is expected to grow from the $200 million it is today to more than $25 billion by 2015. This is what the world needs now—so this is where opportunity lies. We aren't talking about some consumer fad. The creation and support and spread of clean energy industries will happen, as it's just beginning to happen, on such a massive scale that it will transform economies. We're not just talking about making better cars; we're talking about changing the ways we work and the way we live.

For U.S. manufacturers, policymakers, and consumers, this should be a call to arms. A renaissance is within reach. If Americans are the ones who design and build these new technologies, it will re-energize commerce in the United States, creating, without a doubt, millions of high-paying jobs in high-value, advanced manufacturing, and millions more as supply chains extend and benefits ripple throughout the economy.

The New New

As I've said, we know that some manufacturing jobs that have gone offshore—the ones driven mainly by labor costs—aren't coming back. But again, my goal is not to revive obsolete jobs. It's to transform the manufacturing sector for the future. The energy opportunity makes clear that manufacturing has never been more relevant than it is today—or can be tomorrow.

The opportunity is America's to lose. The strengths it will take to seize the opportunity are strengths this nation has in abundance: mastery in science and engineering, an entrepreneurial spirit, and an unrelenting desire to lead the world.

So it shouldn't surprise us that there are already a lot of exciting projects happening in the United States right now. As I mentioned, Dow is excited about what we're doing in Michigan and elsewhere. Other companies, too, are investing in energy solutions. In recent years, most of the major breakthroughs in renewable technologies have been the result of American R&D. By 2010, the United States had about 35 gigawatts of installed wind power, more than any other country.

But if we are really honest with ourselves, we have to face the fact that other countries are more serious about leading the world in clean energy than the United States is. China and Germany, for example, are aggressively promoting clean energy industries in their countries—and are already reaping the rewards. They have given themselves a head start in creating a thriving advanced manufacturing sector.

It's generally acknowledged that the United States doesn't have an energy policy. The current administration is in the process of trying to develop one, and there are some good ideas on the table.

	Germany	Spain	China	United States
		Markets		
Carbon pollution reduction commitment	40 percent below 1990 by 2020	20 percent below 1990 by 2020	40 to 45 percent decrease in carbon intensity by 2020	No binding national policy, although a political commitment to 17 percent below 2005 by 2020
National renewable electricity standard	20 percent by 2020	30 percent by 2020, with carveouts for specific technologies	15 percent nonfossil energy by 2020, with additional policies for specific technologies that effectively strengthen the goal	No comparable national policy, although 29 states have mandatory RES policies and six more states have goals for renewable energy use
National energy efficiency plan	E.U. goal of reducing energy use to 20 percent below business-as-usual projections by 2020. Building codes have increased demand for low-energy houses 900 percent from 1999 to 2007.	E.U. goal of reducing energy use to 20 percent below business-as-usual projections by 2020. National plan has already reduced energy intensity by 11 percent from 2004.	20 percent decrease in energy intensity from 2005 to 2010	No comparable national policy, although 21 states have energy efficiency resource standards. The United States will invest $28 billion in efficiency programs as part of American Recovery and Reinvestment Act.
		Financing		
Feed-in tariffs	Tariff targets emerging technologies, with a total subsidy of $4.6 billion	Tariff amount tied to market growth for specific technologies	Tariff is 7 to 9 cents per kwh for wind, with solar moving toward a similar structure	No comparable national policy, although there are a few state and local feed-in tariff experiments
Government-run "Green Bank"	Government-run KfW provides loans and other financing supports for renewable energy and energy efficiency	Multiple programs, including loan programs for specific technologies and support for strategic projects from government run IDAE	Government-run CECIC will have a portfolio of roughly $15 billion in assets consisting of energy efficiency, renewable energy, and pollution control technologies by 2012	No comparable national policy, although DOE's loan guarantee program provides low-cost financing that leverages private capital and DOE-run ARPA-E supports earlier stage innovation

Energy Policy Comparison

SOURCE: http://www.americanprogress.org/aboutus/reuse.html.

	Germany	Spain	China	United States
		Financing (*continued*)		
Tax benefits	Tax incentives for bioenergy and fuel-efficient vehicles, in addition to a generally low corporate tax rate	Tax exemptions for biofuels	Value-added tax reduction for wind generators and value-added tax rebate for raw materials imports used in wind turbine manufacturing	Production Tax Credit for wind and Investment Tax Credit for solar
Other government funds	Market Incentive Program provides $308 million annually in grants to renewable projects	Funding for energy R&D via multiple government institutions (ENCYT, CIEMAT, and CENER)	Multiple technology R&D programs and large equity investments from the state wealth fund	No permanent national policy, although ARRA has $6.3 billion for research, including advanced batteries, carbon capture and storage, and ARPA-E that develops new clean energy technologies
		Infrastructure		
Workforce and manufacturing infrastructure	Provides grants and interest-free loans with goal of reducing number of young adults without vocational training by half by 2015	National renewable energy job-training center has programs for all sectors and skill levels	Strong domestic content laws and incentives to use domestically produced inputs	No permanent national policy on green workforce development, but related programs include $500 million for clean-energy jobs training and "Buy America" provisions in ARRA, the Workforce Investment Act, and the Green Jobs Act
Grid construction and improvements	Coordinating with neighboring countries to build a "supergrid" for offshore wind power	Upgrading grid with new technologies specifically for renewable energy, including use of electric vehicles as a stability tool	Mandates that grid companies must build interconnections for renewable projects and has plans for smart grid by 2020	No permanent national policy, but ARRA includes $17 billion for grants and loans for transmission and smart grid, which will leverage private capital

Energy Policy Comparison (*continued*)

Still, discrete policy ideas—even excellent ones—don't necessarily add up to a national strategy, or what we might call a framework for America's energy future.

There's a lot of discussion in Washington and across this country about the cost of energy. But what concerns me even more than the cost of energy is the *opportunity cost* of lacking an energy policy. America's surprising reluctance to step actively into this space, where there's a lot of exciting manufacturing to be done, and a lot of progress to be made, risks putting the United States at a competitive disadvantage for generations.

It's ironic: energy independence is closer at hand than it could ever be in the age of oil. But America's policy inertia suggests that it seems willing to trade one form of dependency for another—to let other nations build the clean-energy technologies that we will then buy. It's still within America's means to be a world leader in clean energy, but only if we first wake up to what the rest of the world is doing, and to what we must do.

With that aim in mind, I want to describe what China and Germany are doing. Their efforts show just how much opportunity there is in this space, and what governments can do to help spur these industries. China, of course, is an emerging market that is in many ways rewriting the rules for success—so you probably won't be surprised to learn how successful China has become in the renewable energy arena. But Germany, my other example, is a mature western European economy; conventional wisdom argues that economies like Germany's aren't able to achieve this kind of breakaway success. Yet the results speak for themselves. And they tell us, loudly and clearly, just what we'll be giving up if we fail to seize the moment.

Germany's Green Miracle

Through the 1970s, Germany was seen more as an environmental villain than an environmental hero. Lacking sufficient regulatory protection, many of Germany's lakes and rivers were heavily polluted. But in the 1980s, when acid rain began falling on the country, and when a toxic chemical spill turned the Rhine red in color, popular sentiment began to galvanize, and German leaders, to their credit, responded rather quickly to the call to clean up their country.

Around that time, a man named Klaus Tipfer became Germany's first Minister of the Environment. He created a roadmap that, over the decades that followed, would lead Germany toward building a green economy. Tipfer didn't work against manufacturers; he worked with them to promote industry and the environment at the same time. As the head of a Berlin think tank recently reflected, "The idea was to create markets and businesses that profit from higher environmental standards. Another key was to plan long-term and give industry time to adapt."

"It's green policy," agreed the director of the Nature Conservancy of Berlin, "but it's also driven by German economic interests."

In 2000, Germany passed the Renewable Energies Act, which, among other things, created what's called a feed-in tariff. The idea was simple enough: to pay anyone—any company or individual—who sold back renewable energy to the grid. And it paid them at four times the market rate, with an annual return between 5 and 9 percent. The incentive has proved a powerful one. In less than a decade, according to the Center for American Progress, Germany accounted for more than half of the installed solar panels in the world.

More importantly, Germany had more market share in renewable energy than any other country—as much as Britain, France, and Italy combined. Germany overtook Japan as the world's biggest producer of photovoltaic solar cells, and has become the number one exporter of "renewable energy systems." Today, Germany employs more than a million workers in the renewable energy industry, and expects that number to rise substantially in the coming decades.

What has been good for the German environment has been excellent for its economy. Despite the severe global recession of 2009, a record 17 billion Euros were invested in the German renewable energy sector that year. According to the Center for American Progress, Germany's global export share for wind towers and turbines was above 70 percent in 2006. Its solar cell export share was around 30 percent, and climbing. Roland Berger, a business consulting firm, forecasts that Germany's renewable energy sector will nearly double in size by 2020 and become the country's leading industry.

Imagine that—a renewable energy industry so vibrant that it overtakes the thriving automobile industry responsible for Mercedes and BMW and Volkswagen and Audi. But this is the future. Renewable technologies, as Berger notes, are "expected to leave traditional industries in the dust."

In 2000, Germany set an aggressive goal: Generate 12.5 percent of its electricity from renewable sources by 2012. It met that goal five years early. Its next goal—to achieve 20 percent renewable usage by 2020—will likely be exceeded this year, in 2011. Most economists believe that if Germany stays on track with its clean energy agenda, the country will be able to get nearly half of its power from renewable energy.

These policies have spurred a manufacturing boom in Germany, which in turn, has created high-paying jobs and sustained economic growth. Even at the deepest point of the global downturn, Germany's unemployment rate never exceeded 9 percent. Its economy began recovering from the recession earlier and more fully than its neighbors in the Euro zone.

Germany even has its own clean energy hub, nicknamed Solar Valley. In what used to be Communist-controlled East Berlin, fields and fields of solar panels have risen up among old airfields, surrounded, in part, by old Soviet barracks and other relics of the Cold War.

Here, the solar industry has revived an entire community and done for the region what the automobile industry did for Detroit. More than a third of Germany's total solar production is generated there. The solar manufacturing plants have spawned more traditional manufacturing plants too, those who build key components in the solar energy supply chain. What was once one of Germany's most economically down-trodden areas is quickly becoming the center of its economic growth.

In no uncertain terms, clean energy has become a German way of life. In fact, as of 2010, more Germans were employed in the renewable energy industries than in the coal and nuclear sectors combined. Between 2004 and 2008, the sector increased its workforce by a staggering 73 percent.

And because innovation so often follows manufacturing, German investments in renewable energy have resulted in important new technological breakthroughs. According to *Germany Trade & Invest*, the country's equivalent of the Commerce Department, researchers at the Frauenhofer Institute for Solar Energy Systems recently "achieved a record efficiency level of 41.1 percent for the conversion of sunlight to electricity, a breakthrough which surpassed the previous best mark for efficiency set by U.S. scientists." That kind of breakthrough is especially important for a country like Germany, which is building its solar industry in an often damp and overcast climate (it shares the same latitude as southern Alaska).

New innovations are being tested by researchers and engineers in German facilities every day. And in the meantime, the clean energy manufacturing industry is quickly becoming the German's economic engine, making it one of the most competitive countries in the world.

And it's not alone.

China's Green Revolution

Like Germany, China's entry into the clean energy economy took many by surprise. China's pollution problems are well documented. Many of its major cities—including Beijing and Shanghai—have very low air quality. China's extraordinary economic development over the past two decades—and its unrelenting commitment to its growth strategy—has had side effects like these. The economy has largely been growing on fossil fuels. In 2006 and 2007, China brought a new coal plant online once every week.

China's growth has increased the standard of living of many of its citizens and has helped to build a growing middle class. Of course, with middle class quality of life comes a lifestyle that inevitably requires more energy. It's a one-two punch. China burns fossil fuels to develop its economy, and by succeeding, creates high demand for even more fossil fuels. At the rate it's growing, China's domestic demand for electricity is expected to rise at least 15 percent every year for the foreseeable future. According to the *New York Times*, "China will need to add nearly nine times as much electricity generation capacity as the United States will" just to meet the demand.

In the face of that, China could have continued to do what countries like the United States and Germany had the chance to do for an entire century: use the cheapest energy available, irrespective of emissions, to create sustained economic growth. But China recognized that with such an extraordinary future demand for energy, continuing on its path could literally suffocate the country.

Instead, China decided to rev its manufacturing engines and drive toward a new pursuit: building the world's renewable energy equipment. The plan was aimed at two fronts: continuing China's manufacturing growth by spurring a new industry, and utilizing those renewable technologies to meet China's energy needs.

Some of this has been accomplished in ways that the United States would not wish to follow. China is not a democracy, of course, and the government—though it no longer exerts complete control over the economy—has a tighter hold on the economic reins than the United States ever could—or would. China began, for example, by mandating that 70 percent of all the equipment used in Chinese wind energy projects come from domestic manufacturing. After concerns were raised by other nations, China voluntarily lifted the requirement; but not before jumpstarting its wind energy industry. China also put tariffs on imports of raw materials used in producing wind turbines, and eventually on the turbines themselves.

To make renewable energy more competitive, the government added a small fee to residential electricity bills, then used the proceeds to pay utilities in exchange for purchasing renewable energy for the grid. China also provides a $3 per watt subsidy for solar projects—up front. That means a solar project that adds 20 megawatts of capacity would receive $60 million before the project even broke ground—enough to cover more than half of the capital

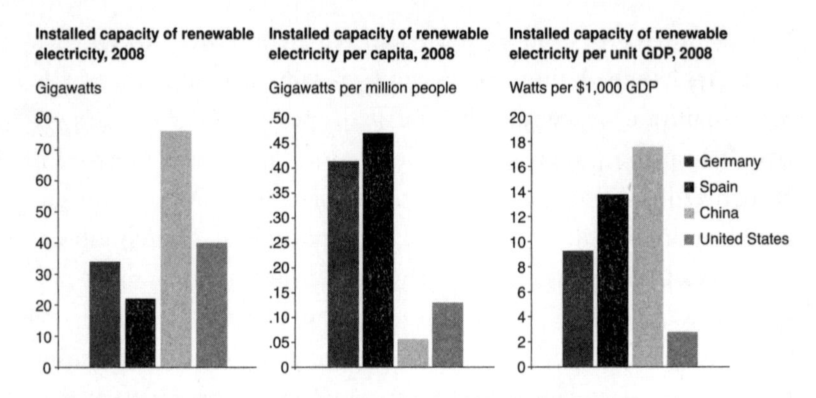

How Much Renewable Electricity, by Country
SOURCE: http://www.americanprogress.org/aboutus/reuse.html.
DATA SOURCE: Center for American Progress, REN21, International Monetary Fund.

costs. This is, by far, the most generous solar subsidy anywhere in the world.

China's relentless pursuit of a clean energy economy has paid off. In 2004, the Chinese government invested just $2.5 billion in the renewable energy sector. In 2009, it invested $34.6 billion, almost twice the amount that the United States spent. China is now a world leader in clean energy manufacturing, and is quickly becoming the world's biggest consumer of renewable energy.

Today, China has more installed renewable energy capacity than any country in the world. It became the world's biggest maker of wind turbines in 2009, and the largest solar panel manufacturer in 2010. It makes a staggering one third of the world's solar panels. Most of those panels are shipped overseas to countries like Germany and the United States. China is also working to ramp up its nuclear energy capacity, with 50 new reactors expected to be built by 2020. According to Thomas Friedman, during that same period "the rest of the world combined might build 15."

The Chinese are aiming high—and increasingly hitting the mark. By 2020, the country is aiming for 150 GW of installed renewable power at home, which could account for as much as 30 percent of China's overall power supply, and would represent more than four times what the United States currently has operating.

In the United States, we can see examples everywhere of just how effective Chinese renewable manufacturing incentives really are. In 2008, Massachusetts sought to attract Evergreen Solar, a solar manufacturer, to build a facility in the state. It provided the company with a $58 million incentive package—including about $20 million in cash grants.

"That was very helpful," Rick Feldt, CEO of Evergreen Solar said at the time, "But if I put it in perspective, it's a $430 million facility. $20 million is about 5 percent."

Less than two years later, Evergreen, faced with the realities of a competitive global market, decided to move the final assembly of its solar panels from the facility built in Devens, Massachusetts, to Wuhan, China. "We're going to China as quickly as we can," he told reporters.

China offered Evergreen an incentive package that Massachusetts couldn't begin to match. "As we go to China," explained Feldt, "we're getting low interest loans on 65% of the factory and equipment . . . You have low labor costs and low overhead costs in China, but you also get enormous help from the government and so it's difficult to compete in the United States if you have to contend with higher labor costs and lower government assistance."

Evergreen is building in Germany, too. "We built our joint venture factories in Germany because it's federal help, not just state help. We got 45 percent on the first factory—not five—and we got 30 percent on the next two factories."

Already a manufacturing heavyweight, with the government's muscle behind the push for a green economy, China will continue to attract multinational corporations to invest. A company called Vestas just built the world's largest wind turbine manufacturing facility in China. "You have to move fast with the market," Jens Tommerup, the president of Vestas China, told the *Los Angeles Times*. "Nobody has ever seen such fast development in the wind market."

And as with other industries, China's success in renewable manufacturing has attracted R&D centers, too. Innovation, once again, is following production. The world's largest privately funded solar research facility began operation in 2010 in Xi'an, China. In late 2009, China announced that 16 new energy-related R&D centers would be built, focusing primarily on wind and nuclear power, as well as improved efficiency.

Though they went about it in a different way, the Chinese, much like the Germans, saw the immensely attractive opportunity that renewable energy represents and are taking action to seize it. Both nations are now poised to lead the clean energy industry over the coming decades, to benefit from the advanced manufacturing jobs it provides, and to enjoy the resulting economic growth.

They are also poised to do something else: leave the United States behind.

America Can't Compete

Again, I want to acknowledge that America is getting some of this right, more so here than in most other areas of manufacturing.

At the federal level, the United States has provided renewable energy tax credits, which have helped make the market for wind and solar power more viable. President Obama's stimulus package offers grants to companies that install solar energy systems, and the administration has removed the cap on a substantial tax credit for homeowners who install solar panels.

States, too, are working hard to attract renewable energy manufacturers. Oregon gave SolarWorld, a German-based company, $40 million in business tax credits to attract the company to build in Hillsboro. The state is also trying this year to copy the German model, testing the use of a feed-in tariff that will pay those who sell renewable energy to the grid 10 times the market rate.

California has also established a feed-in tariff, and has made a strong push on both the wind and solar fronts. Between 2007 and 2008, it doubled the amount of solar power installed in the state. And Texas, which gets 5 percent of its energy from wind power, is considering putting incentives in place to attract solar to the state. Luke Bellsynder, executive director of the Texas Association of Manufacturers, has said that solar advocates "tell us constantly that panel manufacturer X or Y was looking to Texas, but decided to go to other states because they have a better [renewable portfolio standard] or better incentives for production."

Competition among states helps the nation. And because many global producers believe the United States has the potential to become the largest market for renewable technologies, the country is, in some cases, actually attracting foreign investment.

But this piece of good news should not overshadow the reality that the United States is facing. America is falling behind other countries in building a clean energy industry—and forfeiting its best possible opportunity for growth in the manufacturing sector.

It is true that the federal government has offered incentives for renewable energy. But the incentives in the stimulus are temporary and oversubscribed. Most solar producers that would like to build a plant in the United States have been told that certain tax credits are no longer available. The future of others seems uncertain.

And while many states should be commended for the incentives they are offering, there is no way that a single state—most of which are facing crippling budget crises—could ever compete dollar-for-dollar with economic powers like China and Germany. Oregon may have been successful in attracting a SolarWorld facility, but more than two-thirds of the revenue that the company generates is still from Germany. "No one can compete with the German feed-in tariff," says Boris Klebensberger, the head of the American branch of SolarWorld. "You can see the success. You implement good political policies and it works."

The biggest challenge facing the United States is that, unlike countries like China and Germany, the United States still lacks a coherent, comprehensive national energy policy. For the most part, political leaders have left the development of a clean energy industry almost entirely to market forces, save for piecemeal initiatives. Unlike China, Germany, and others, the United States still has no targeted policy on carbon emissions. Unlike China, Germany, and others, the United States still has no national commitment to meeting a certain percentage of its energy needs from renewable sources. Nor does the United States offer the kind of clear incentives other countries do to meet those goals.

As Kevin Parker, global head of Deutsche Asset Management Division, told Reuters, "You just throw your hands up and say . . . we're going to take our money elsewhere."

These failures send a signal to companies looking to invest: they say the United States is uncertain about how much it values renewable energy and how willing it is to support investments over the long term. In the United States, a power company is presented with a choice: it can buy renewable energy equipment and invest in a market without clear government backing, or it can take the hint, shrug its shoulders, and continue to operate the coal- and natural gas-fired power plants that are already paid for and offer more reliable returns.

On this front, as on others, inaction by U.S. policymakers compounds the competitive disadvantages our economy already faces. In January 2010, Thomas Friedman wrote about an e-mail he received from the CEO of eSolar, a California-based start-up. In January 2010, the CEO announced in Beijing "the biggest solar-thermal deal ever. It's a 2 gigawatt, $5 billion deal to build plants in China," he wrote, "using our California-based technology. China is being even more aggressive than the U.S. We applied for a [U.S. Department of Energy] loan for a 92 megawatt project in New Mexico, and in less time than it took them to do stage 1 of the application review, China signs, approves, and is ready to begin construction this year on a 20 times bigger project!"

Make no mistake about it: broadly speaking, this is a perilous turn of events. In 2009, China surpassed the United States in renewable energy investment, according to the Pew Charitable Trusts. So did nine other countries. America hasn't been the world's leading solar manufacturer since the 1990s, and today makes fewer than 5 percent of the world's solar panels. Only one of the top 10 solar producers in the world is located in the United States. Only one of the top five wind turbine manufacturers, GE, is based here.

Consider this alarming statistic: according to the Center for American Progress, "over 70 percent of all component parts for installed clean and efficient energy systems in this country are currently imported—many from countries with higher labor costs and environmental standards than ours."

Chinese cars are as much as one-third more fuel efficient than cars made in the United States. General Motors is building hybrid cars—in China. As a percentage of GDP, China is investing 10 times more money in renewable energy than the United States.

These numbers are devastating. While the United States cedes the industries of the future to other nations, most policymakers seem not to notice. If America doesn't shift gears quickly and take advantage of the energy opportunity, we are almost certain to trade a present where we import oil from Saudi Arabia and Kuwait for a future where we import solar panels from China and wind turbines from Europe.

We have occasionally heard the right rhetoric from our politicians. In 2009, President Obama told an audience, "We can cede the race for the twenty-first century, or we can embrace the reality that our competitors already have: The nation that leads the world in creating a new clean energy economy will be the nation that leads the twenty-first century global economy." But as the president himself would probably agree, that vision has yet to be met with substantive action—with tangible, permanent policy changes that would get America's economic engines moving again.

There are plenty of important economic reasons for the United States to follow China and Germany's lead to spur its domestic clean energy industry. But there are other reasons—of basic morality—that deserve consideration, too. The Energy Information Administration reports that the manufacturing sector was responsible for 1.4 billion metric tons of carbon dioxide emissions in 2002. That's nearly 20 percent of all U.S. carbon emissions. Putting a green energy plan in place would allow the manufacturing sector to take responsibility for building a cleaner, more sustainable future. It's not just good economic policy. It's the right thing to do.

Chapter 5

Building Tomorrow

When companies like Dow talk about R&D, we use the term pipeline. The choice of metaphor is not accidental. Pipelines are long; sometimes they're circuitous. They carry something essential, but can take a while to deliver it. Businesses that depend on R&D don't just ask themselves, "What's the next big thing?" We ask, "What's the next big thing that *comes after* the next big thing?" Science and technology rarely yield results overnight. Investments might not pay dividends for years—even decades.

So it is with the work of nations. Running a country, of course, isn't much like running a company, but a few common principles apply. In both cases, you will face crises that demand immediate attention; you will face pressure—whether from shareholders or voters—to show results right away, and your response can't be "wait a generation or two." At the same time, you will also need to stay focused on the long term and resist expediency. You'll need to put good ideas in the pipeline, and make investments that might not bear fruit until your successor is in charge. In other words, you've got to deal with reality as it is—while you shape reality as it can and should be.

This is something most businesses struggle with. Partly, it's the result of the competing pressures of the ownership structure, of capital markets that favor quick gains. As capital moved into hedge funds and other short-term investment vehicles, shareholders have come to expect big returns quarter over quarter, which, in turn, drives companies to focus on achieving those gains, sometimes at the expense of the long-term. It is also, frankly, the result of uncertainty, of businesses deferring some of these decisions while they grapple with a policy environment that has too many unknowns.

I've tried to take the long-term approach to my work at Dow—never more so than in 2008 and 2009, when we, like so many other companies, were hit hard by the economic downturn. Our earnings and equity earnings both dropped. Our operating rates hit historic lows. Then, to make matters worse, a would-be partner pulled out of a major joint-venture deal at the last minute, despite having signed a binding contract requiring them to close.

We were at a crossroads. Some people outside the company urged us to abandon our long-term strategy. Step back. Scale back. Walk away. Instead we went forward. Why? Because we looked within ourselves and concluded we had the right, long-term strategy for our company. Despite the crisis in the credit markets, despite the collapse of our market cap, we knew we had to see that strategy through. We closed the deal, and took a big step from what we were able to do today to what we are capable of doing tomorrow. We were betting big on our future.

That same tension—between short-term and long-term thinking—is apparent in our public life. Especially in times of economic distress, it's tempting for politicians to focus on the present at the expense of the future. The pressures of the *permanent campaign* compound the problem. Politicians seek—and are rewarded for—the quick fix, the silver bullet. Major societal challenges that lack easy, expedient answers are neglected. These are the issues where actions we take today likely won't show real results for 10 or 20 years, sometimes more. These, therefore, are the issues most often ignored by political leaders—pushed off to the next Congress, the next president, the next generation.

In this chapter, I want to discuss two of those issues: our education system and our infrastructure. Both are crucial to our national prosperity. Both are essential for our competitiveness. And for all the recent attention given to both, neither system is remotely ready to help us meet the challenges and seize the opportunities of the twenty-first century.

Education: "A Permanent National Recession"

You could fill a book with the issues that need to be addressed in American education—and many have. That is not my intention here.

My goal, with this chapter as with the others, is to focus on policy failures that impair our national competiveness. I want to discuss education as it relates to our economy. I want to talk about what's at stake for the manufacturing sector—and, more specifically, what kind of consequences we can expect as a nation should we fail to adapt.

Some of this, we've heard before. Read just about any article about education, listen to just about any speech by an education reform advocate, and one of the first things you'll hear is that American students are falling woefully behind their international peers. We're told, for example, that in 2006 American 15-year olds ranked twenty-fifth out of 30 nations in math and twenty-fourth in science. Our elementary school students actually do a bit better than that in the international rankings, which would be good news—if it didn't suggest that the longer students stay in the American education system, the more that system fails them.

It's reasonable to ask, why should we care? Does it really matter than the average student in Japan or Finland is better at math than an average American student? Sure, it wounds our national pride, but does it have actual, measurable consequences? And if so, what are they?

A recent study by the consulting firm McKinsey & Company offers a stunning answer: If the United States had closed the educational achievement gap between its students and those of better-performing nations in 2008, our GDP would have been $1.3 trillion to $2.3 trillion higher that year. That's the equivalent of 9 to 16 percent of GDP. To put that in perspective, today's manufacturing sector makes up just 11 percent of GDP.

So closing the gap between nations would create a substantial amount of new wealth in the United States. It would also close dangerous gaps within America, because that resulting prosperity would be more widely shared—between traditionally low-performing and high-performing states, between people from rich families and people from poor ones, and between whites and minorities. McKinsey found that closing the performance gap between states would have resulted in GDP being as much as 5 percentage points higher. Close it among rich and poor and GDP goes up another 3 to 5 percentage points. Close it between races—GDP goes up another 2 to 4 percent.

These numbers are staggering in their implications. There isn't a single other policy issue that has such a dramatic impact on America's economic performance. As the McKinsey study notes, "the persistence of these educational achievement gaps imposes on the United States the economic equivalent of a permanent national recession."

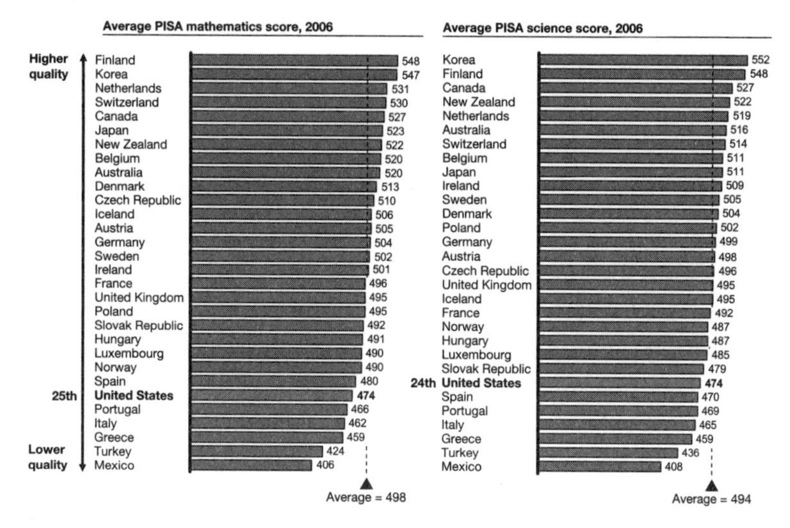

The United States Trails Other Countries

Source: Exhibit 1 from "The Economic Impact of the Achievement Gap in America's Schools," McKinsey & Company, Social Sector Office.

Data source: Organisation for Economic Co-operation and Development.

How did we get here?

Didn't the United States used to have the world's greatest educational system? Indeed, during the 1950s and 1960s, American schools were considered among the best in the world. As recently as 1995, the United States was still tied for the top spot in graduation rates. But by 2006, we had dropped from first to fourteenth. That's partly because many other countries are investing substantially in education, raising their standards, and often exceeding them. They have longer school days, longer school years, and give students homework over the summer. American students spend 40 percent less time studying than they did 50 years ago. That's not how it is abroad. Other countries are far more serious about getting results.

We sometimes console ourselves that while our schools—we can't deny it—are failing the most disadvantaged students, our star pupils can beat star pupils anywhere in the world. That's just not so anymore. Other countries' best students are doing better than our best students. According to McKinsey, Korea, Switzerland, Belgium, Finland, and the Czech Republic produce at least five times the proportion of top performers in their student populations as the United States.

The U.S. education system, as Bill Gates has said of high schools in particular, has become "obsolete." This can't be disputed. For every 100 ninth graders we have, only 18 will go on to earn an Associate's or Bachelor's degree in a timely fashion. If I ran my plants with an 82 percent failure rate, I'd be out of business in a matter of days. Too many students, having done all that was required of them, are graduating from high school unprepared either for college or the working world. It's not essential, in my view, that every one of these students goes on to a four-year college. But every student needs to get a degree that has actual value and meaning in the workplace.

Developing the Right Skills for the New Workplace

The workplace, after all, is changing dramatically. In 1973, half of all manufacturing workers never finished high school. Only 8 percent had any education beyond high school. It was a very different time, and manufacturing, as I've described, was a very different sector. Today, fewer low-skilled jobs are available—and even fewer are being created. According to a study by Georgetown University, 63 percent of jobs in the next four years will require more than a high school degree. And yet we have a system that is continually churning out unskilled, unprepared high school graduates (assuming they even graduate), sending them out into an economy that has no place for them. The outcome is tragic for them as individuals, and deeply worrying for our national future.

That isn't to say that there aren't some exciting things happening in American education right now. Among them are charter schools operated by innovative organizations like KIPP and Green Dot that are narrowing the achievement gap among urban low-income populations. There remain questions about whether these programs can be scaled up, but for now they are offering students chances that were nonexistent a generation ago. We also see more manufacturing institutes forming, new approaches that show promise, and new philosophies that get impressive results. There are schools that use new technology to try innovative new teaching methods. Khan Academy, for example, uses a library of more than 1,600 videos to teach students anywhere in the world. There are also traditional schools that are excelling. Among them are Midland High School and H.H. Dow High School in Midland, Michigan, which succeed in part because of our company's support and presence on campus. In nearly every school district—even the lowest performing—we can point to at least one success story.

The problem is that these successes are surrounded by failures. Because we value—as Americans always have—local autonomy and cultural differences, we have created an education system that isn't really a system at all; it's a loose confederation. It's a patchwork, with substantial variations in standards, quality, and results. These vary not only state by state, but also district by district, school by school, even classroom by classroom within the same school. As the McKinsey study notes, "This confirms what intuition would suggest and research has indicated: differences in public policies, system-wide strategies, school site leadership, teaching practice, and perhaps other systemic investments can fundamentally influence student achievement."

The countries that have built the world's best education systems have succeeded because they have departed from this distinctly American approach. A *New York Times* editorial explained that "the countries that have left the United States behind in math and science education have one thing in common: They offer the same high education standards—often the same curriculum—from one end of the nation to the other." Contrast that with the United States, where a child's education, as the editorial put it, "depends primarily on ZIP code."

Teachers also often lack relevant training in the subject matters they teach. Few chemistry teachers, for example, have a chemistry background. Instead, teachers are encouraged to earn master's degrees in generic fields like "education" that have little relationship to what they will spend their careers teaching.

Preventing a Worker Shortage

A properly educated workforce is an issue of deep concern to business, especially to manufacturers. A plant that employs 300 people at an average wage rate of $50,000, for example, will require, over the course of a decade, a human capital investment of $300 million. We cannot afford to get an investment of that magnitude wrong. Neither can we afford to locate our plants where we know we will face a serious skill shortage. A recent report from the World Economic Forum said exactly what so many CEOs are thinking these days: "Human capital will soon rival—and may even surpass—financial capital as the critical economic engine of the future. The scope of the challenge is so broad that no single stakeholder can solve it alone."

Several years ago, a collection of prominent business groups, including ones in which I am a member, put out a report warning of a coming shortage of American talent. "This is on the top three CEO agendas of every company I know," William Green, CEO of Accenture, told the *San Francisco Chronicle*. Our worry isn't just that our children are being educated poorly; it's that they're being educated poorly in the subjects most relevant to our economic well-being. There isn't enough focus on science, technology, engineering, or math in our schools (known as STEM in education policy circles). The country will need 400,000 new graduates in STEM fields by 2015; but we won't get them. And when we don't, companies in search of new talent will have no choice but to search elsewhere.

Baiju R. Shah, the chief executive of a nonprofit that's trying to turn Cleveland into a medical device manufacturing hub, expressed frustration about the applicant pool to the *New York Times*. "That's where you see the pain point. The people that are out of work just don't match the types of jobs that are here, open and growing."

Ben Venue Laboratories, a pharmaceutical manufacturer, told the *New York Times* that the company had reviewed 3,600 job applications in 2010 and only found 47 people to hire for 100 positions. The workers didn't even need to have advanced degrees. All that was required was to pass the most basic skill test, with the bar set at a ninth-grade math level. But a substantial portion of the applicant pool failed. "You would think in tough economic times that you would have your pick of people," said Ben Venue's CEO. You'd think. But you'd be wrong.

I can't say I was surprised when a 2009 survey by the Manufacturing Institute found that one out of three manufacturers reported "moderate to serious" skills shortages. I hear this concern from friends all across the manufacturing sector. Companies are desperately seeking top-level engineers and employees who have the dexterity to manage enormous projects across multiple borders. But too often, the pool of workers that fits those criteria is far too small. Dow experiences these challenges firsthand. Over the next five years, we may need to replace as much as 30 percent of our global workforce. Yet there is a significant shortage of chemical engineers in this country—so much so that the highest paid college graduates in the United States are now the relative few with chemical engineering degrees.

Over time, we can expect these problems to become more pronounced. As more and more Baby Boomers reach retirement age, the lack of young Americans with the skills to go into advanced manufacturing is going to exacerbate an already severe problem. In Ohio, 30 percent of the state's manufacturing workers will be eligible to retire in the next five years. As of 2008, one in four STEM degree holders in the work force was 50 or older. When these people leave the workforce, we won't have a strong pool to replace them. We can expect a race for qualified workers over the next few years. And when organizations remain in the United States despite not having found the talent they seek, they will be forced to settle for workers they know are subpar, and hope—through a combination of luck and training—that they will be able to perform. Inevitably, that will have a significant impact in terms of lost production and capital diverted away from innovation and R&D.

The aerospace industry expects to take that hit especially hard. Northrop Grumman estimates that about 60,000 workers—half of its total workforce—will be eligible for retirement within the next decade. Fifteen percent of Boeing's engineers are currently at retirement age. Lockheed Martin could lose half its workforce, too. Lockheed CEO Robert Stevens expects that over the next decade, the company will need to hire 142,000 engineers—yet U.S. colleges are producing only about 60,000 new engineers a year. As Stevens has written, "The looming tech talent shortfall will have an impact far beyond any single firm or sector. Science and engineering aren't just crucial for national security; they're critical for economic growth."

The effect of this shortage is substantial. As the president of the Federal Reserve Bank of Minneapolis recently noted, if we could find a good match between the jobs that are open and the skill sets available from our workers, the U.S. unemployment rate would be two points lower than it currently is. That represents about three million jobs.

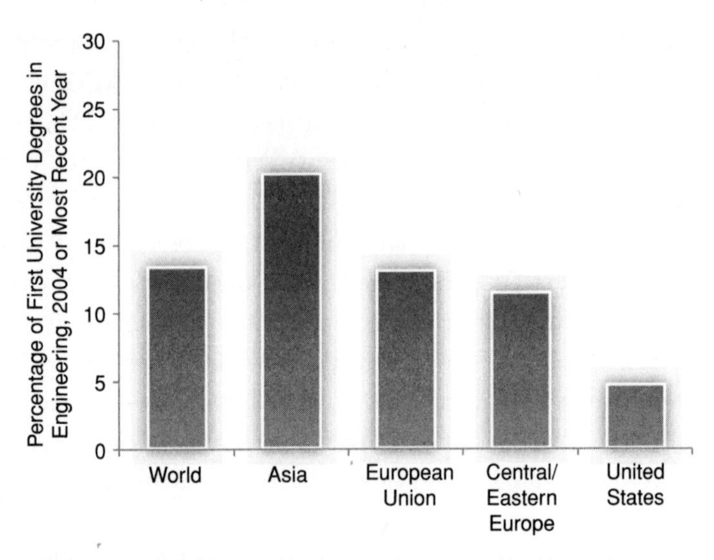

Percentage of Engineers around the World
SOURCE: Copyright © 2009 by The Manufacturing Institute. *The Facts about Modern Manufacturing, 8th ed.* 2009.
DATA SOURCE: National Science Foundation, *Science and Engineering Indicators*, 2008.

What America Doesn't Understand That Other Nations Do

With the stakes this high, I'm often asked why I suppose it is that America is graduating so few students in such important fields while other countries are graduating so many. In China, India, Germany, and Japan, for example, engineering is seen as one of the most rewarding and prestigious careers one could possibly find. Graduates in those fields are highly respected—and compensated accordingly. Why, then, don't Americans see these fields as exciting, ennobling—worth pursuing?

My opinion is that it's a function of what Americans choose to value as a society, and how we articulate those values, both in our curricula and our culture.

Just off of Route 2 in Eastlake, Ohio, is a facility owned and operated by Astro Manufacturing and Design. The company makes everything from high-tech medical devices to aerospace and military equipment. Frequently, Astro provides tours of its facility to students—as a way of getting the region's young and malleable minds excited about a future in manufacturing. It is not an easy task.

"I rarely get anyone who has any interest in manufacturing," explains Astro's tour leader. Students arrive at the facility with the preconceived notion that manufacturing is "boring and monotonous," that it's mindless work, the kind they associate with the assembly line.

I hear that all the time. Americans are used to thinking about manufacturing jobs as a caricature of what it was decades ago: jobs that required relatively little skill and even less critical thinking. They imagine themselves soldering the same metal part to the same metal frame hour after hour, day after day, without deviation. They see jobs like this as a dead end, as an anachronism, destined for outsourcing—or oblivion. It's no wonder, then, that for every Ph.D. in physical sciences or engineering, America graduates 18 new lawyers and 50 new MBAs.

This is a chronic problem and it exists, even among the children of engineers. I frequently hold town halls with Dow employees and their families. They can tour our facilities, learn about our processes. When I ask how many of those kids are studying science and engineering, only a few hands go up. And these are the children of engineers! I suppose I shouldn't be surprised. I pushed science at the dinner table with my kids from a very young age, but none of them ended up going into engineering. Even when parents value these careers, they are operating in a society that encourages kids to look elsewhere.

Sometimes I wonder whether some other countries—namely, the ones that missed out on the heyday of the industrial age—are better able than we are to see manufacturing as it is today, rather than what it used to be. They don't have long-held preconceptions, as we do. When they think of manufacturing, they don't think of their fathers leaving for the steel plant in the morning. They see high-paying, skill-intensive jobs. High-tech jobs that are always evolving—jobs operating incredibly complex machines, drawing or reading complicated blueprints, mixing specialized chemicals, perfecting microchips at the molecular level. Jobs that require an excellent education. So the education systems in these countries are designed to create more of what they seek, value, and admire.

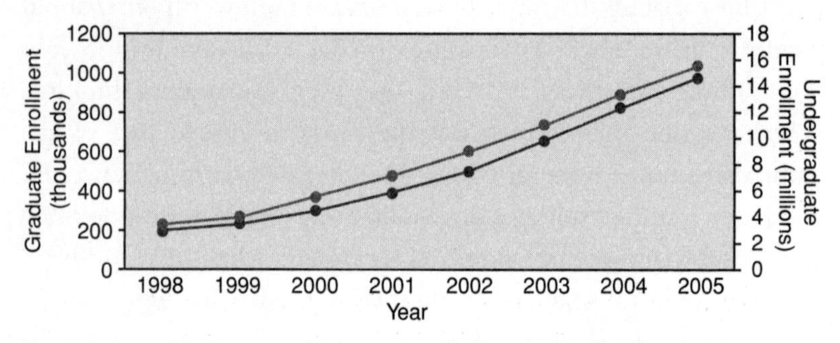

China Student Enrollment at University Increases
SOURCE: Reprinted with permission from "China's 15-year science and technology plan." Copyright © 2006, American Institute of Physics.

It's worth nothing that the talent shortage hurting the United States is also affecting nations globally. The number of Chinese employers who reported difficulty finding skilled workers rose 25 percent this past year. But unlike the United States, these countries are taking serious actions to solve this problem. When China entered the world stage as a major economic power, it resolved to supply, over time, enough highly skilled workers to meet its own demand. Nicholas Kristof of the *New York Times* cites the example of Dongguan, one of many Chinese manufacturing boomtowns. Twenty years ago, there was not a single college in Dongguan. Since then, the city has invested a fifth of its budget in education and built four universities. Today, incredibly, 58 percent of Dongguan's college-age residents attend a university. Country-wide, China has increased its undergraduate enrollment six-fold since 1998.

India, too, places enormous value on education, and is making great strides. Though much of the country's state-run education system remains underfunded and inadequate, India's Seven Institutes of Technology are highly regarded, seriously competitive, and well funded, too. *U.S. News & World Report* ranks the Indian Institute of Technology Bombay among the top 30 engineering institutions in the world, outranking at least four Ivy League universities.

The Tortoise and the Hare

It is important to remember that the United States still leads the world in innovation—because it still leads the world in higher education. *The Christian Science Monitor* notes that "U.S. scientists still publish twice as many of the most influential research papers as their European counterparts, and four times as many as a group of countries [called] the Asian 10, which includes China and India." But those nations and others are investing billions in closing the gap.

Their determination—and our unwillingness to think, as they do, a generation ahead—makes it likely that the high-tech, high value-add products of the future will be imprinted with a cruel rebuke to the United States: "Designed *and* built in China." Or Germany. Or someplace else.

Education isn't an end in itself. It has a purpose—many purposes, in fact. Beyond empowering our children to think critically, beyond inspiring intellectual curiosity, beyond encouraging social mobility, our schools exist to prepare the next generation for the world they will be entering—and the economy they will be inheriting. It is failing in that vital purpose.

Our education system is indeed, a pipeline. We will get from it what we put into it, a generation down the line. This is a problem that craves action, as quickly as we can provide it. It is also a problem that, thankfully, has many new, innovative and promising solutions. In Chapter 7, I will lay out a policy agenda for education that I believe will move us toward the global competitiveness we need.

We must also recognize that, even if we succeed in confronting our education crisis, it is not the only long-term challenge we face. Our global competitiveness doesn't just require a well-educated public. It doesn't just require a population that can match the job market. It also requires a national infrastructure that can meet the needs of a changing economy.

A New Foundation of Infrastructure

It's hard for us to imagine George Washington as the father of American infrastructure. It isn't central to the legend of Washington—the great general, the man who refused to be king. But indeed, George Washington was fascinated by infrastructure, and determined to push the United States to build. Even before America's declaration of independence, Washington advocated the establishment of the Potomac Canal Company, which was aimed at connecting the Potomac to the Ohio and Mississippi rivers. In 1783, he traveled to New York to survey a potential route for a canal connecting the Hudson to Lake Erie. During that trip, Washington wrote a letter in which, on reflecting on the potential of American infrastructure, he said, "I could not help taking a more contemplative and extensive view of the vast inland navigation of the United States . . . and could not but be struck with the immense diffusion and importance of it." From Washington's perspective, these innovations in infrastructure were not just about moving people. They were about moving goods and raw materials, about building an economy that could sustain itself. Washington understood that the success of the American experiment would depend, in no small part, upon a steady flow of commerce—the lifeblood of any nation.

It was that basic understanding that helped drive the nation's westward expansion through the nineteenth century. It was that understanding that created transcontinental railroads and bridges, and drove the nation to build immense projects aimed at connecting industries across the country.

During the twentieth century, America built on those strides to create the greatest infrastructure the world had ever known. Franklin Roosevelt's New Deal programs put the country to work building roads and bridges, hydroelectric dams, power plants, sewage systems, pipelines, irrigation systems, levees, and more. These were, as Roosevelt described, "new foundations for business," investments that could spur and sustain our national prosperity—as they did indeed. Roosevelt's administration also invested in air transport, working with private airlines to create air traffic control systems, to improve radio communications, and to maintain safety standards. Subsequent presidents built on these foundations. When Eisenhower became president, he signed the Federal Aid Highway Act, and became known as the father of the interstate system. Kennedy and Johnson, too, expanded and upgraded our infrastructure, investing billions in transportation.

That national commitment played an indispensable role in making the United States the world's supreme economic power in the twentieth century. It enabled the manufacturing sector to grow and prosper, connecting businesses to consumers, within the country and around the world.

We have conquered the seas, mastered the skies, and covered our continent with a remarkably extensive network of roads and rail lines. Short of teleporting goods from here to there, it's hard to imagine a means of transport that we haven't come up with already. Given that, what does a twenty-first-century infrastructure look like? Is it any different, really, than what we've had in the past, or what we have today?

The answer is yes, very different. Imagine if every acre of windy land and every potential solar farm in America was connected to population centers with energy-efficient transmission lines. Imagine a smart grid that brings power to homes and businesses while using 20 percent less electricity. Imagine an advanced network of the highest-speed broadband cables ever produced that connect the country, rural and urban areas alike. Imagine if the United States had highly automated, state-of-the-art ports from which exports could leave our shores at a highly efficient pace. Imagine high-speed rail corridors, where people and products could be propelled at 200 miles an hour toward their destinations, while burning less than half as much carbon as an airplane or truck over the same distance.

I can tell you what all this would mean for Dow: we would save tens of millions of dollars a year in energy costs. We would save millions more in shipping costs. We would have easier access to emerging global markets. But that's just the start. We would also have the opportunity to enter a new, potentially lucrative market: designing and producing the innovations that would make this new infrastructure work.

For example, much of this modern infrastructure would utilize wireless sensors that could collect and transmit data from roads, from pipes, from waterways. As the *New York Times* noted in April 2009, "Computer-enhanced infrastructure promises to be a lucrative market." Already, a number of American manufacturing companies are moving into this space, and doing incredibly exciting things. The *Times* reports that IBM is working on a number of "smart grid" programs, utilizing computer sensors and software to maintain power lines and reduce energy usage. One such project found that peak loads could be reduced by 15 percent. That's the equivalent of eliminating 30 coal-fired plants over two decades. It's all part of their *Smarter Planet* initiative, which aims to use technology to solve some of the world's most vexing infrastructure problems.

IBM's microchip factory in Vermont has tested this sensor technology to improve the efficiency of industrial water use. (Manufacturing companies use incredible amounts of water in the production process for cooling and steam generation—a semiconductor plant can use as much water as a small city.) According to the *Times*, the plant was able to reduce its water consumption by 27 percent (that's 20 million gallons a year), even while increasing manufacturing output by 30 percent. Clearly, this is a win-win for the company: it saves a great deal of money while producing a high-value-add product that, when scaled, should be quite profitable.

That's why manufacturers are so interested in an upgraded infrastructure. It benefits the sector in three key ways: it can create a market for the manufacturing of products that are part of the infrastructure system. It can allow us to operate more efficiently and cost-effectively. And it can help our products to reach deeper into emerging markets for less. That trifecta, when considered across the entire economy, will result, without question, in an acceleration of growth.

That's the infrastructure we need. But it's not the infrastructure we've got.

In the several decades since the post-war building boom, the flow of commerce has expanded, the pace of our national life has sped up considerably, and the demands we place on our infrastructure have increased exponentially. Under these stresses, FDR's "new foundation for business" is cracking—and showing its age.

According to the American Society of Civil Engineers, one in four of the nation's bridges is structurally deficient or functionally obsolete. One in three of our major roads is in "poor or mediocre" condition. As economist Paul Krugman wrote, in a number of states, "local governments are breaking up roads they can no longer afford to maintain, and returning them to gravel."

Without major highway expansions—and absent a serious commitment to public transportation—Americans have become stuck in the slow lane. We spend, on average, 4.2 billion hours a year in traffic—which costs the economy an estimated $78 billion a year in productivity losses. Piecemeal repairs to roads and highways—the policy equivalent of a finger in the dyke—cost taxpayers $67 billion a year.

The nation's pipes are so old—and in such disrepair—that an estimated seven billion gallons of clean drinking water leak out of them every single day. Systems meant to clean wastewater that's part of the manufacturing process are so outdated that they, too, leak billions of gallons a year—into U.S. surface waters.

America's ports—crucial for international trade—are in disrepair. In fact, over the past 20 years, we've added just one major export terminal. These ports can often be an impediment to effective export. Older ports that haven't been modernized can't move goods through the system nearly as efficiently as their international counterparts. Those delays add real costs to manufacturers.

A trip along America's inland waterways, meanwhile, is like a voyage back in time. The locks on our rivers are incredibly outmoded. Of the 257 locks in operation, more than a third are 10 years past their intended lifespan. Some are crumbling, and many simply aren't designed to meet modern needs. Most barge tows on the Mississippi River, for example, are 1,200 feet long, but many of the locks are only half that. At each lock, therefore, the tows must disassemble, pass through in pieces, and reassemble on the other side. That takes time and costs money.

Our electricity demands have increased by 25 percent in the past 20 years. But our electricity infrastructure has not been updated to meet that demand. Broadband Internet was invented in the United States—but we've now fallen behind other countries in terms of quality, speed, and deployment of the technology.

It's hard to hear these staggering statistics and not wonder: How is this possible? How could this have happened?

I believe it is largely because most Americans—certainly most American politicians—take our infrastructure for granted. We drive on highways that previous generations built, through tunnels and over bridges constructed in another century, and many of them have become such landmarks, such emblems of American life, that we assume they have a kind of permanence, irrespective of what we do or don't do. Other than civil engineers, few of us think of, say, a highway overpass or a suspension bridge as a delicately balanced piece of technology that must be carefully maintained—or, increasingly, replaced.

That is, until tragedy strikes, as it did in 2007, when an interstate bridge collapsed in Minnesota, killing 13 people and injuring 145 more. Then we stop and think. But only for a moment. Of course we want our roads and bridges to be safe. And if you asked the American people whether they want modern ports, waterways, and power systems, they'd say, yes, absolutely. But outside the state or local level, you'd be hard pressed to find a constituency that's energized about the issue. And infrastructure, fundamentally, is a national challenge. We don't prioritize it—and our elected leaders don't, either. The federal budget reflects that lack of commitment. We dedicate just over 2 percent of the budget to infrastructure—a 20 percent decline relative to GDP since 1959, when President Eisenhower built the highways on which we still drive.

Funding the Future

In Congress, transportation bills do pass, and lots of money gets spent. But there's no grand design. Roads get repaved, but no strategy emerges. President Obama's Recovery Act has made substantial investments in infrastructure—but the principal focus, for very good reasons, has been job creation in the near term. That puts a premium on *shovel-ready* projects that can begin construction immediately. That has been the standard: if a project is ready to go, it receives federal funding, whether or not the investment contributes to a broader vision of a twenty-first century infrastructure.

This is not to say the money's been squandered. Some of it, at least, has been well spent, in ways important to our future: $7 billion has been allocated to expanding broadband, for example, and $3.4 billion will go toward updating the country's electricity grid to better meet demand. But these projects don't automatically add up to a more productive, efficient, and modern infrastructure. It's a bit like randomly putting dots on a piece of paper and hoping that when you connect them, they'll look like something recognizable.

The Recovery Act, in any event, isn't representative of the way infrastructure projects get funded. The typical process is worse. Congress doesn't appropriate transportation or infrastructure dollars based on their expected rates of return on national objectives. It doesn't require projects to compete for finite resources. Instead, lawmakers rely almost entirely on a politicized, haphazard, and counterproductive process. Projects get funded through earmarking, based on the relative clout of members of Congress, and on lobbying by interest groups. The result is that the money we do spend on infrastructure is often wasted. Rather than targeting our tax dollars where they can have the greatest impact, Congress, in the words of Bruce Katz, founding director of the Metropolitan Policy Program at the Brookings Institution, spreads that money around "like peanut butter"—in a thin and even coating.

I understand how business gets done in Washington. I'm not going to say we can eliminate every "bridge to nowhere." But I will say we can't let that infamous bridge in Wasilla, Alaska become a metaphor for our entire approach to infrastructure in this country. This is too serious to be determined by pork-barrel policymaking. Neither can shovel ready be our standard for everything. Near-term job creation is important—but long-term job creation is even more so. The latter, and the prosperity that flows from it, will depend on the speed and volume and reliability of our intrastate, interstate, and international commerce. We can't let the twenty-first century be the era of the bottleneck, or the breakdown.

Other nations see this imperative and are meeting it. They are investing—substantially—in modernizing their roads, transit systems, and electrical grids.

- China, for example, has invested billions in building transmission lines across the vast countryside and is already outspending the United States on a smart grid. Over the coming years, China will be criss-crossed with new highways and rail lines.
- Both China and Europe are upgrading their major ports and hubs to keep pace with the expansion in international trade.
- Brazil recently launched an $878 billion program to upgrade its national infrastructure—that's a full $100 billion more than the entire American stimulus package.
- Japan has long had the most advanced high-speed rail system in the world, and continues to maintain its edge—while capitalizing on its expertise by exporting its high-speed technology.
- Australia has implemented a plan to run fiber-optic broadband cable to 93 percent of the country's homes, spending tens of billions of dollars to connect its citizens to each other, and to the rest of the world.

Each of these nations has developed a long-term strategy and is taking determined steps toward fulfilling it. They don't see a new infrastructure system as the ultimate goal; the objective, in the end, is leadership in the global economy. A state-of-the-art infrastructure is, quite literally, a high-speed pathway toward that future.

I recognize that much of what I'm describing isn't cheap. No doubt the federal deficit—and the politics of deficits—will hold the United States back from investing in infrastructure on a scale that matches other nations. As a businessman, I have no trouble understanding that.

At the same time, I think we have to be frank with ourselves, not just about what it will cost to upgrade our infrastructure, but what it will cost if we fail to do so. There are, to be sure, creative ways to do this affordably and responsibly, ways that can accomplish our aims without spending more than we are able to. We need to prioritize infrastructure investments the way we prioritize defense investments. We need to do more than just create weapons of war; we must also create the instruments of peace—roads, bridges, grids, and telecommunications. I will share my specific ideas about how to move forward with a twenty-first-century infrastructure agenda in Chapter 7. But again, I think it's critical to understand that the nature of a sound long-term investment—like the dollars Dow invests in its R&D pipeline—is that over time, it will bring real returns.

Businessmen like me don't expect guarantees. Risk is inherent in everything we do. But I think it's a safe bet that if American leaders can sketch out a vision of what the country needs to look like in the twenty-first century—if we draw up and stick to a national blueprint—then our tax dollars can be targeted accordingly, and the test of any individual project will not be whether it's shovel-ready, but whether it readies us all for the opportunities ahead.

Chapter 6

Built to Compete

American history is dotted with turning points, moments in which our actions and choices have led us, irrevocably, in a new direction. From our founding to the Civil War, from the Great Depression to the Cold War, this nation has responded to overwhelming challenges with a grit and resolve that makes me proud to call America my home.

I firmly believe that our capacity to compete—and win—in a global economy is the next great challenge we face. I have no doubt that 10, 20, or 30 years from now, when we look back at the expanse of history that will have unfolded, we will see this moment as a turning point, the real start of the twenty-first century—the moment when we either took up the challenge and brought America fully into the global economy, or instead, let indecisiveness and inaction unravel our prosperity, and with it, our quality of life.

I didn't write this book just to call attention to this crisis and its causes. I wrote it because I believe this is a crisis we can solve. I'm an Australian of Greek heritage who has lived the American dream. I know that the free-market capitalism that built this nation has taken it further down the democratic path than the Greeks could ever have imagined. But I know, too, that like the Greeks, like the Romans, we face the real possibility of decline.

Over the next two chapters I will lay out a comprehensive agenda with a single objective in mind: to reinvent the American manufacturing sector, and revive the American economy.

The agenda is anchored in a basic premise that might sound counter to what you would expect from a CEO: We need government action. We need a national economic framework. And we need it now.

For decades, many Americans, especially business leaders, have rejected the very idea of government intervention in our economy, believing that the best way to create long-term economic growth is to let the markets—and the markets alone—rule. But we've learned especially painfully in recent years that for all the wisdom of the markets—something in which I still believe—there are important areas where markets, frankly, aren't so wise. Markets cannot put a price on certain things we nonetheless value deeply. They cannot always ensure that consumers are protected or that the economic foundation we are building is solid.

Too often over the past few decades, we have assumed that our actions as individual companies—each acting on behalf of its own customers and shareholders—would add up, in the aggregate, to a coherent and successful economic model. But that faith has not been justified by events. As we have seen, markets can never be a substitute for the kind of thorough, long-range, strategic thinking that is the responsibility of government. Every citizen (and corporate citizen) must have a voice in the discussion, but in the end, only governments can orient a nation toward its essential goals. Only governments can create a policy climate that encourages—or in our case, discourages—innovation and commerce. Only governments can set a vision for our future. Only governments can create the kind of framework that will move markets to respond.

This has always been true, but never more so than today, now that we live in a truly global marketplace where other countries are competing for the same jobs, the same dollars, the same economic growth and all that flows from it. Whether we think it's fair practice or sound policy, we know that other governments are intervening on behalf of their nations. They are almost single-mindedly dedicated to creating jobs within their borders—and inviting companies like mine to do it for them.

The United States isn't going to stop that from happening. We wouldn't want to even if we could. But because it is happening, we face a choice. We can either get in the game, and play to win—or, alternately, decide that we're content to be the only major nation that plays by its own rules, no matter how outmoded and outmatched those rules may be.

That, to me, doesn't feel like much of a choice at all. There's no one in this country who wants to watch helplessly as our standard of living steadily erodes. But that's what we're doing every day that we hold onto the fiction that classic, laissez-faire capitalism is the only alternative to the kind of total government control of the economy we associate with places like China and Singapore. It isn't.

Of course, we aren't China. Neither are we Europe or anyplace else. America has its own form of government, its own system of beliefs, its own enduring values—the legacy of this nation's founders. The U.S. government is never going to exercise the kind of control over the economy, or engage in the level of central planning, that some other countries do—nor should we, nor must we. There is, just as there has always been, a uniquely American formula for success. Our challenge today is to define what it is.

Let's remember we're not starting from scratch. The American manufacturing model, the one that created so much prosperity here over the past century, that created the middle class, was the most successful economic model in history. Other nations learned from that success. Now we can learn from theirs—and apply those lessons in a uniquely American way.

An Ambitious Agenda

The agenda I'll lay out is ambitious. Seeing it through to the president's desk will not be easy. While Washington can sometimes get big things done, we all know that in general, it is incredibly hard to get even small things accomplished. A supermajority in the Senate is required—and even that isn't always enough. Still, we cannot allow the flaws in our political system—and the poison in our political climate—to limit our national imagination. We need to think big. We don't have any other choice. The scale of the challenge demands it.

We must not only think big, but think holistically. One of the things I hope I've gotten across in the preceding pages is that our challenges are interconnected. They therefore defy piecemeal solutions. What's true in business is true in government as well:

when you try to address sweeping problems in scattershot fashion, you leave gaps. You fail to see connections. You end up with a basketful of ideas—many of them good ones—that may or may not bear some relation to each other, and most likely don't fix the problem at hand.

First and foremost, we need a strategy—a national economic framework. As the cliché goes, you have to know where you want to go before you can figure out how to get there. You need a roadmap. Again, as we have seen overseas, economic strategy isn't the accidental outcome of uncoordinated actions. It is the product of leadership—at the highest levels of government—and partnership with key stakeholders, especially the private sector. Where this happens, the strategy is clear, and becomes a guide to every subsequent decision. It enables policymakers to ask, "Does this serve our strategy?" and not simply, "Is this, in itself, a good policy?"

I know this is a tall order. I'm not under any illusion that somebody is going to pick up this book, look at my proposed solutions, and turn them into an omnibus bill that passes a few weeks later. I'd be delighted if that happened, of course, but I'm a realist. I am also a pragmatist, and I see no need for this all to be done in one fell swoop. There's nothing wrong with taking incremental steps toward grand goals. After all, America didn't go to the moon in a single shot.

What follows is a policy agenda for the economic framework I believe we must follow. It aims to accomplish five key objectives:

1. Make it easier for businesses to keep or locate their operations in the United States.
2. Remake the manufacturing sector with a focus on advanced, high-value products.

3. Create an economy that can sustain itself, and can, in turn, produce long-term job creation and economic growth.
4. Prepare the next generation's workforce for the changing economy.
5. Improve America's global competitiveness, both in the short and long term.

This chapter is the first part of that agenda. I call it the *Immediate Impact Agenda*. These are policy solutions that, if implemented, would have an almost immediate effect on the manufacturing sector and on the country's ability to compete globally. In the next chapter, I will put forth some policies that I see as foundations for long-term, sustainable growth.

Changing the Way We Tax

The U.S. tax system was designed in an era when American companies faced relatively little competition from abroad. One of the most critical ways that the U.S. government can revive the manufacturing sector is to reform the tax code that's strangling it. Many well-minded political leaders have argued for common-sense reforms. Senators Judd Gregg and Ron Wyden have authored a bill known as the Bipartisan Tax Reform and Simplification Act that should serve as a model of the kind of reform I envision. "This investment-oriented proposal will. . .encourage people to create jobs and make our nation more competitive," noted Gregg, upon introducing the bill in the Senate.

As I discussed in Chapter 3, many countries are lowering their corporate tax rates in an effort to attract multinational corporations to invest within their borders. Yet over the past 25 years, the United

States has actually *raised* its rate, and now has the second highest in the world. The United States needs to take swift action to make its tax rates more competitive.

At the same time, out of concern about the federal deficit, we should make these tax code changes in narrow, targeted ways that serve key national goals—in this case, the goal of attracting investment in advanced manufacturing. Rather than cut the corporate tax rate across the board, we should create a new—and substantial—manufacturing tax credit that reduces the effective rate on manufacturing companies considerably.

We should also look beyond the corporate tax rate to other tax policies that would benefit the manufacturing sector. The R&D Tax Credit, for example, is essential for businesses in the United States. But it is temporary, and needs to be reauthorized every year. In 2010, the R&D Tax Credit expired and has yet to be reauthorized. This threatens America's long-term innovation capacity, discourages businesses from investing in R&D domestically, and prevents businesses from planning long-term R&D budgets. It creates enormous uncertainty.

The United States needs to remove that uncertainty by making the R&D Tax Credit permanent. But it needs to do more than that. It needs to expand and simplify the tax credit, given the importance of investment in basic research and development. In the fall of 2010, President Obama, to his credit, proposed these kinds of changes to the R&D credit. But given the political climate in the run-up to the midterm elections, there was little hope of its passage. It's important, in the coming Congress, that the president work with both parties to turn this proposal into legislation, and get it to his desk. If we are going to remain the world's leading innovators, our policies must articulate and encourage that goal. We shouldn't be twenty-third in the world in tax treatment of R&D investments. We should be number one.

National Incentive Strategy

In order to attract multinational business to their shores, countries aren't just lowering their tax rates. They are also providing all sorts of incentives—everything from free land to low-interest loans to contract prices on feedstocks. Some companies have even been offered the money to pay up to 80 percent of workers' salaries for as much as a decade.

In the United States, incentives like these are often criticized as tantamount to corporate welfare, and, perhaps, appropriately so. Corporations are making plenty of profits, the argument goes, and don't need any more giveaways from governments. The problem is that if we refuse to offer these kinds of incentive packages while other countries are aggressively outdoing one another, we put America at a clear competitive disadvantage. The price of inaction on this front is, as a result, far greater than the price associated with leveling the playing field.

Simply put, we need to start competing for the investments that businesses make and the jobs they create. At times, that will require strategic national investments made to attract specific businesses—and industries—to build in America rather than abroad.

Accepting that reality is one thing. Identifying a way to implement it is quite another. It isn't hard to imagine a scenario in which the only businesses getting incentive packages from the federal government are those who hire a team of lobbyists to do their bidding. It could also become the case that a patchwork of incentives attract businesses but don't add up to a coherent strategy. That would obviously still be of some value— to the businesses receiving the incentive packages and to the Americans who would get the jobs. But an ad hoc approach would almost certainly deplete federal resources while doing little to revive the manufacturing sector.

But that doesn't mean we should give up altogether on the idea of providing strategic incentives. Instead, Congress should create and fund an independent entity—called, say, the National Economic Growth Bank—that would offer incentives to businesses in growth industries, businesses with a particular ability to create sustainable jobs and sustainable growth. The Bank should be controlled by a bipartisan board, appointed by the president, and should include representatives from industry, as well as economists, scientists, and engineers.

Regulatory Policy

On March 3, 1993, a little more than a month after taking office, President Clinton took the podium at the White House to launch an initiative that, until that point, was unheard of in our national dialogue. He announced the formation of a "national performance review," intended to "make the entire federal government both less expensive and more efficient, and to change the culture of our national bureaucracy away from complacency and entitlement and toward initiative and empowerment. We intend," he continued, "to redesign, to reinvent, to reinvigorate the entire national government." Vice President Gore would spearhead the effort, and the administration would demand results.

The idea was simple enough: the federal government, as those of us in business are well aware, is slow-moving and slow-acting. Regulations often conflict with each other. Bureaucracy—the proverbial red tape—weighs down new enterprise instead of encouraging it.

Six months later, the president's performance review had identified tens of thousands of places where the government could reduce its size without reducing its efficacy, where it could increase its speed without compromising its oversight. And to the surprise of many, including myself, the president and Congress spent the next seven years working aggressively to implement the recommendations.

They eliminated more than 16,000 pages of unnecessary federal regulations. They rewrote 31,000 pages of regulations into plain, easy-to-understand language. They worked with key agencies, including the Environmental Protection Agency (EPA), the Food and Drug Administration (FDA), and the Occupational Safety and Health Administration (OSHA), to ensure that the rules being written were focused on key performance metrics rather than process measures. It was a critical initiative aimed at making business and government work better together—and it was quite successful.

But since the end of the Clinton administration, the mantle of reinventing government has not been taken up again. It was not a priority of the Bush administration, nor has it become one for the Obama administration. Absent a consistent effort, absent a government dedicated to weeding out its own inefficiencies, the problem of bad regulations perpetuates itself, leaving us, more by default than by design, with overregulation.

I hope that the work that began under President Clinton will begin again in earnest. Congress should pass new legislation requiring a new performance review and swift action to implement its recommendations. That review, I believe, should be conducted with six principles of reinvention in mind:

1. **Harmonize and simplify rules.** Throughout the federal bureaucracy, thousands of rules are unnecessarily complex and contradictory. And there are plenty of rules that no longer serve their original purpose. Any initiative aimed at streamlining the federal government must start with simplifying and harmonizing the rules across agencies and cabinet departments.
2. **Enact performance standard regulations.** Traditionally, regulations are written to be prescriptive. They set specifications that must be followed, covering everything from what materials can be used by manufacturers to how they can be used—and when. What these regulations fail to explain is why they exist in the first place. What is the ultimate purpose they are serving? Research has shown that the most successful regulations are those that are written in a way that defines the end they are trying to achieve, rather than the means to get there. Instead of forcing businesses to follow rigid guidelines, they create flexibility, allowing businesses to work creatively to efficiently achieve the same end. The goal of the regulation is achieved more quickly and at a lower cost for business—and the government. Reorienting regulations to focus on these kinds of performance standards is a critical step toward making government work better.
3. **Increase collaboration with business.** Agencies that regulate industries must obviously maintain a degree of independence from them. But that doesn't mean they can't talk with businesses to gain a better understanding of what is achievable. By working with industry in the initial, regulatory development phase, agencies can create standards that are more easily implemented and even more effective.
4. **Accelerate permitting decisions and extend permit durations.** The review and approval processes for operating

permits can take an exceptionally long time, often forcing a business or industry to miss a key window of opportunity. The government needs to rewrite regulations in a way that will accelerate the permitting decisions process. It can also reduce the amount of delay by reducing the number of times a business must apply and reapply for a permit. Extending the duration of permits is, obviously, the easiest and quickest way to accomplish that.

5. **Enhance benefits of self-reporting.** In 1995, the Environmental Protection Agency established a new kind of audit policy. Rather than having the government participate in a police-style oversight of companies, the EPA created a system of self-reporting. Companies would audit themselves and report violations, in exchange for substantially reduced penalties and other mitigations. This may sound gutturally unwise. After all, without strict outside oversight, you would expect to have events like the BP oil spill become commonplace. But as it turns out, the self-auditing policy has actually been extremely effective. And you don't have to take my word for it. EPA directors of both Democratic and Republican administrations have boasted about the effectiveness of the audit program. First, it's cheaper for a company—which knows its own operations better than an outside entity—to do the audit. But most importantly, the incentive of reduced penalties is substantial and, indeed, motivating. Companies have participated in rigorous self-audits, and have, as a result, improved compliance with minimal cost to the EPA. This concept works. And it should be expanded to agencies beyond the EPA.
6. **Benchmark our regulations against those of other developed countries.** When new requirements are considered by agencies, they should be benchmarked against equivalent

regulations of our major developed country competitors. The United States should be willing to write regulations that are more strict, and at times, more costly than those of other countries. But only when necessary. Regulators should be required to consider—even if they then choose to disregard—the impact that new regulations would have on the competitive capacity of the industries they regulate.

Everyone Needs Good Trading Partners

The American public has turned sour on free trade agreements, a feeling stoked by those in Washington (on both sides of the aisle) who argue that free trade has done little more than hasten the decline of U.S. manufacturing. Nothing could be further from the truth. As I previously mentioned, America enjoys a trade surplus with countries with whom we share free trade agreements. When other countries lower their tariffs, it makes our exports more competitive and gives us access to new markets. That boosts demand for our products, increases our jobs, and boosts our economic growth.

Also, at this stage it should be obvious that a free trade agreement is not a prerequisite for a company to move its operations offshore. Plenty of factories have closed in the United States and opened in countries with whom we do not have any free trade agreement at all. There are, as I've described extensively, plenty of incentives for companies to move offshore that far outweigh concerns about tariffs.

In the meantime, our ambivalence about trade is hurting the domestic manufacturing sector. Protectionist policies, aimed at keeping manufacturers in the United States, don't actually work.

We frequently hear talk of penalizing companies for moving offshore, but this is more punitive than productive. As I've argued, we should be luring companies to stay in the United States, rather than trying (and surely failing) to force them to do so. By opening new markets for our exports, we'll be giving companies one more good reason for building products in the United States.

So free trade agreements deserve a dedicated push. Rather than politicizing the issue, our elected leaders should take an honest look at what free trade agreements will do for their constituents. A great example is the free trade agreement between the United States and South Korea, which has been languishing in Congress for far too long. South Korea is already our seventh largest trading partner. In 2008, the United States imported $48.1 billion in goods from South Korea, and exported $34.7 billion. That resulted in a trade deficit of just over $13 billion. However, the U.S. International Trade Commission estimates that if we ratify the free trade agreement, it would add $10 to $12 billion to U.S. GDP, and an additional $10 billion in merchandise exports. That's nearly an 80 percent reduction in our trade deficit with Korea. And it would mean, without question, new jobs and new growth in the United States.

Free trade agreements with Colombia and Panama—which were approved by those nations back in 2007—have remained stalled in Congress, as well. Again, these deals would be a massive boost for U.S. manufacturers. Look at Dow. If the free trade agreement with Colombia passed, it would immediately eliminate 90 percent of Colombian tariffs on more than $300 million of U.S.-made Dow products that we currently export there. Reducing those tariffs would save Dow $22 million annually, and would increase the demand for our products in Colombia.

We have to reject the misguided and dangerous rhetoric that paints trade as the enemy to economic growth. It isn't the enemy. On the contrary, it enables growth. I therefore hope that Congress and the President will take the following steps:

1. **Ratify the Colombia, Panama, and South Korea free trade agreements.** This can—and should—be done swiftly. The sooner the agreements are ratified, the sooner we can begin reducing our trade deficit.
2. **Enter negotiations with our other major trading partners to create free trade agreements that are mutually beneficial.** There are a number of opportunities around the world for the United States to enter bilateral negotiations to establish free trade agreements. We should seize these opportunities.
3. **Work with the World Trade Organization to successfully conclude a commercially meaningful Doha Round.** In 2001, more than 100 countries came together in Doha, Qatar to negotiate a historic trade deal that would lower global trade barriers and improve market access. A decade later, they're still negotiating. Again and again, trade talks have broken down, usually on a developing/developed country divide. And each time the parties have returned to the table, they have failed to make any measurable progress. If the United States can exercise the leadership necessary to successfully negotiate a meaningful agreement through Doha, it would reduce tariffs all over the world and, in doing so, dramatically boost our export demand. Success at Doha could be the world's most effective stimulus package.

★ ★ ★

As you can see, this is an ambitious agenda. But it isn't pie-in-the-sky. There isn't a single piece of it that should be beyond our reach. Each policy proposal is reasonable, and targeted specifically at reviving the U.S. manufacturing sector.

What would happen if this agenda was adopted in full by the United States?

We would become instantly and dramatically more competitive around the world. Companies that are currently considering closing their operations in the United States and moving offshore would have compelling reasons to stay. New startups that have created innovative products and are ready to scale up their operations would find the United States a much more attractive place to put down roots for the long term.

Our exports would increase dramatically. As a result, our trade deficit would narrow substantially. This would stimulate economic growth and, most importantly, create sustainable, good-paying jobs. In the fight for global competitiveness, the United States would emerge again as a force to be reckoned with—as a major player that has no plans to forfeit its position as the world's greatest economy.

Chapter 7

The Long Game

I was only 10 years old when President Kennedy was assassinated. I was living in Darwin, Australia then, truly a world away from that terrible event. But I can still remember watching the news as the tragedy unfolded, seeing an entire nation in mourning. When I eventually made my way to the United States, I continued to be fascinated by Kennedy, with his ability to connect with the American people and with what it meant to be American. I have since read a great many of his speeches.

One in particular always sticks out in my mind. On March 23, 1962, the President went to the University of California at

Berkeley and spoke of the "momentous events around the world": of the Cold War, the threat of nuclear annihilation, and the need for wisdom grounded in "the long view."

Kennedy ended his speech with a peroration about the need to confront the most complex and long-term challenges we face. "We must think and act not only for the moment, but for our time," he proclaimed. "I am reminded of the story of the great French Marshal Lyautey, who once asked his gardener to plant a tree. The gardener objected that the tree was slow-growing and would not reach maturity for a hundred years. The Marshal replied, 'In that case, there is no time to lose, plant it this afternoon.'"

It is hard for a nation to look far down the road when there are obstacles, or crises, in its immediate path. But as Kennedy suggested, the seeds of future greatness must be planted well in advance. Conversely, when we ignore problems because they seem too great or too distant, they only become more difficult to solve.

Much of this book has focused on the problems facing manufacturers right now. But we cannot ignore issues of equal importance that have a much longer time horizon. They might seem less urgent—but believe me, they're not. Addressing them today is the only way to ensure that America will have the capacity to reach for—and achieve—sustained economic prosperity tomorrow.

What follows is the longer-term portion of the national economic agenda I hope America will pursue. Some of the solutions I describe here can have an impact on our economy in five years or fewer, but my primary focus in this chapter is on those that might take 10, 20, or more years to bear fruit. But again, as President Kennedy understood, these are trees that must be planted today.

The Human Element: Education and Immigration

Of all the potential competitive advantages we could build in the twenty-first century, none is more important than human capital—what we at Dow call *the human element*. The country that has the best educated, most talented, most competitive workforce will surely lead the world in the generations ahead. With some serious reforms to our education and immigration systems, I believe the United States can be that nation.

A New Look at Education

In 2010 Bill Gates gave a speech to the American Federation of Teachers and spent some time discussing the work of his foundation. He began by acknowledging the skepticism that exists around the idea that our education system can ever be reformed.

> The United States has been struggling for decades to improve our public schools. We have tried reform after reform. We've poured in new investments. Since 1973, we have doubled per-pupil spending. We've moved from one adult for every 14 students to one adult for every 8 students.

But, he continued,

> Despite these efforts, our high school scores in math and reading are flat. Our graduation rates have plunged from second in the world to sixteenth. And our 15 year olds now rank behind 22 countries in science and 31 countries in math. There's no denying it—these are dismal results in student achievement.

Much of what America has tried in education reform has failed. Much of what we thought would work hasn't. For years, that undeniable reality has led political leaders—and the public—to conclude that America's education crisis can never be solved.

Today, however, that despair is lifting just a bit, being replaced by a growing—if cautious—optimism. Thanks to hard-working reformers and a number of organizations, not least the Gates Foundation, we finally have the research to tell us what is and what is not working. We have a better sense than ever of what we need to do.

Because manufacturing companies are in such desperate need for highly skilled workers with backgrounds in science, technology, engineering, and math (STEM), much of the education agenda that follows is focused specifically on improvements in those subject areas. I wouldn't begin to suggest that I know how to fix the endemic problems that exist across the educational spectrum. But I do try to consider STEM in its larger context. For here, as elsewhere, the nation needs a comprehensive strategy for reform, in which individual efforts reinforce one another—setting in motion a virtuous cycle of success.

K–12 Reforms

- **Teacher quality.** For a long time, experts thought the key to improving education was to reduce class sizes. The argument, which seemed reasonable on its face, was that if students could get more personal attention from teachers, performance would substantially improve. For years, school districts across America have prioritized class-size reduction. But to our collective disappointment, students haven't done much better as a result.

New evidence points us in a more fruitful direction. We have learned that the most decisive factor in student achievement isn't the size of the classroom, but the quality of the teacher in front of it. Students do better, it turns out, when 30 of them are taught by an excellent teacher, than when 15 are taught by a mediocre one. As Bill Gates has noted, "when each of the variables under a school's control is correlated with student achievement, the teacher is the one that makes the biggest difference—and that difference can be dramatic."

We therefore need to focus our reform efforts on getting better teachers in our classrooms. Toward that end, we need better methods to evaluate teachers so that we can reward the good ones and, if necessary, improve or remove the bad ones. The era of entrenchment must end.

We also have to make sure our teachers have the tools to teach better. They need constructive feedback, partnership, and support. And most importantly, they need training in the subjects they teach. Right now a number of states increase the salaries of teachers who hold a master's degree, and offer subsidies for teachers who want to earn one. But the vast majority of these degrees are in fields like education with no direct bearing on the subject areas the teachers will actually teach. Students suffer as a result. Evidence suggests that a master's in education has no positive impact on the quality of the teacher who holds it. We need to stop encouraging current and future teachers to pursue such degrees, and instead reward them for earning degrees in areas that are relevant to their jobs.

This is especially important in STEM fields, where mastery of the subject matter is absolutely essential. Believe me, there's no bluffing your way through a chemistry curriculum (though many have tried)! To bring more STEM-educated teachers into our classrooms, there should be scholarships for students who pursue STEM-related majors—and who pledge to teach in that field.

- **National standards.** America can no longer afford to have a patchwork system of education, in which standards differ from state to state. The countries with the best education systems have all implemented uniform national standards. You cannot truly gauge how well teachers are teaching and how well students are learning unless you have meaningful, consistent, and high standards by which to evaluate them.

  Recently, we have seen real progress on this front. Race to the Top, one of President Obama's most innovative initiatives, pits states against each other in a competition for education funding. For the first time, states have to make substantial changes to their education systems *before* they receive the funding. That includes adoption of national educational standards. These, it's important to note, were not created by Washington. Instead, state leaders joined together to develop what's known as the Common Core standards. According to a July 2010 *New York Times* article, 27 states quickly adopted the Common Core, and another dozen were expected to do the same. "I'm ecstatic," said Education Secretary Arne Duncan. Standards, he added, have long been "the third rail of education, and the fact that you're now seeing half the nation decide that it's the right thing to do is a game-changer."

  Still, we cannot be satisfied until every state in the union signs on. Any state that fails to do so will undoubtedly be left behind. So President Obama should continue to use Race to the Top to persuade the remaining states to join the effort. And he should continue to use this model to push states to achieve the education reforms we know they can.

- **Innovation in classrooms.** The education system should encourage—and reward—innovation in teaching methods, particularly in STEM subjects. Many American students struggle with science and math early on, and become permanently discouraged from pursuing those subjects. New teaching models and new technological tools can make a difference. Teachers should be given the flexibility to experiment. Research can identify best practices, which can then be shared and spread.

We must also look beyond K–12 for reform opportunities. The stronger our K–12 system becomes, the more students will choose to continue their education. But whether high-school graduates go on to college or move directly into the workforce, we need to equip them with the tools to excel. It's the only way to fulfill the American ideal of equal opportunity—and to ensure the nation has enough skilled workers in critical fields.

Continuing Education Reform

- **Skills training.** Through the 1980s and 1990s, skills training was a top priority for the United States. Many states initiated programs through community colleges and started customized training programs, aimed at creating a talent pool from which local companies, especially manufacturers, could draw. Both the first Bush administration and the Clinton administration emphasized the importance of such programs to keep the workforce competitive.

  But as James Moore notes in *Manufacturing a Better America*, this approach ended abruptly as globalization took hold and manufacturing jobs began disappearing from the United States. Priorities shifted—and so did these programs. No longer would they prepare workers for careers in manufacturing. Instead, they would retrain displaced workers for jobs outside the plant. While this made sense in certain respects, it was a powerful sign that the United States was, in effect, giving up on manufacturing. "The new era of worker training is short-term and concentrates on resume writing and techniques for quickly finding a new job," writes Moore. "It is aimed at getting workers off of public unemployment insurance benefits as quickly as possible."

Today, when Americans graduate from high school, there are simply too few options for skills-based continuing education. It is time for the government to recognize once again that skills training in manufacturing is crucial for the nation's global competitiveness. The United States should reinvest in programs that can prepare Americans for high-skilled, high-paying jobs in advanced manufacturing.

- **Provide tax credits to employers who help their eligible employees pursue continuing education.** Of course, the private sector has a role as well. Companies have a vital interest in developing their own programs and helping their workers to continue to learn and build skills. Dow does this quite successfully. Through partnership with Delta College, a community college in Midland, we help employees earn associates degrees in chemical process technology. Other companies, like GE, have similar programs. The United States can encourage even more to follow suit by providing tax credits to companies that provide substantive skills-based training.
- **STEM scholarships.** Generous scholarships should be provided for students who seek a bachelor's degree or a postgraduate fellowship in a STEM field. What we fund as a nation says a great deal about what we value—and I can think of few things of greater importance to our national future than creating the next generation of scientists, engineers, and innovators.

From K–12 to higher education and continuing education, it is time to see reform not just as a priority, but a real possibility. We know the steps we need to take to upgrade our educational system for the twenty-first century. What we need is a commitment to do it.

A Nation of Immigrants

We in America tend to view immigration through a narrow lens. Most of the national discussion—you could also call it a shouting match—concerns the influx of low-skilled, undocumented workers. This is a concern that requires a thoughtful solution. But my focus here is on a different pool of immigrants: highly skilled workers who study in our universities, who work for a while in our labs, and whose temporary visas require them to leave our country.

Every year, the United States provides student visas and temporary work visas to some of the best and the brightest from all over the world. But in very few cases do we allow them to stay here for long—which prevents them from innovating for our companies, starting businesses of their own, and investing for the long term.

Perhaps that's due to the belief that we should be "saving" American jobs for American workers. But keep in mind: more than 1 million jobs in science and technology will open up *this year alone*. But only 200,000 new graduates will have the skills to fill them. It will be years before the education reforms I described above will have a measurable impact; between now and then, we've got to fill the gap of skilled workers. We can't do it when only 65,000 temporary H1-B visas are being issued annually. That's not nearly enough. I can tell you from experience, there is nothing more frustrating than having to turn away applicants with all the right skills. We need to raise that cap and extend their time in the United States. Or they will, of course, go elsewhere.

As an immigrant myself, it is my great hope that the United States will change its immigration policy to make these shores more hospitable for entrepreneurs. Consider this: between 1995 and 2005, one out of every four U.S. tech companies was established by at least one foreign-born cofounder. That includes Google, eBay, Intel, and Yahoo. We should make it easier for brilliant women and men to move to the United States, and build their futures—and ours—here.

Innovation and Competitiveness

There are a number of short-term ways we can improve our innovative capacity and our global competitiveness; I outlined some in the previous chapter. What follows here are solutions that will prepare us not just for short-term job creation, but for long-term, sustainable growth.

Energy

Nowhere is there more opportunity for a manufacturing revival than in the emerging renewable energy sector. To make the most of this opportunity, our political leaders need to treat energy not just as an environmental or a national security issue, but as an economic issue—and an economic opportunity.

A new energy policy could dramatically boost advanced manufacturing if it addresses four key areas: It should

1. Create a high demand for renewable products and technologies in the marketplace.

2. Ensure that domestic energy sources can meet that new demand.
3. Increase the incentives for energy-related innovation.
4. Reduce the staggering cost of energy for manufacturing companies.

Creating a Market. As I've mentioned before, Dow is incredibly excited about the solar shingles we are manufacturing in Midland, Michigan. We are enthusiastic about entering such an important market, one that is expected to be about $5 billion a year. That's nothing to scoff at.

But imagine for a moment that Congress passed a new policy, similar to Germany's, allowing citizens who equip their homes with solar shingles to sell the excess power they're generating back to the grid. And imagine what would happen if the United States, like Germany, subsidized that sale, so that Americans would receive more than market value for the energy they sold.

There would suddenly be enormous demand for solar shingles in this country, not just on new homes, but on old ones as well. The policy would make it more affordable for owners of existing homes to make their homes more energy efficient. If anyone doubts that Americans respond—in big numbers—to incentives like these, look no further than the cash for clunkers program that encouraged people to replace their old, inefficient, polluting cars.

I'm confident that if the government does for homes what it did for cars, millions more Americans would want solar shingles on their roofs; they would want to get into the business not just of consuming power, but of producing and selling it. Instead of a $5 billion market, we might be looking at a $50 billion market. Instead of creating a few thousand jobs, we could create hundreds of thousands of jobs dedicated to meeting that rising demand. We'd be building a whole new industry.

If the United States is going to lead the world in renewable energy, it must help to define the market here at home. We should follow the German model by implementing these same kinds of incentive programs. California and Oregon have already introduced a similar policy with positive results. This is the most promising way to accomplish our goal. A study from the National Renewable Energy Laboratory found that these types of incentive programs (known as feed-in tariffs) were responsible for 75 percent of the world's solar deployment, and 45 percent of its wind deployment. The bottom line is that if you give the American people an incentive to buy renewable technologies, the American people will buy them.

Building an Industry. Of course, stimulating demand is a critical first step toward building a vibrant renewable energy sector. But it is just a first step. Government policies must also reduce risk and uncertainty for manufacturers in this area.

I'm often asked, given the uncertainty around renewables, how Dow has managed to jump with both feet into the business of solar and battery technologies. The answer is simple. We had help. There was far too much risk, far too many unknowns, for us to enter the market to this degree. But the federal government and the state of Michigan came to us and provided the direct financial support we needed to offset the high risk of our initial investments. That removed enough of our uncertainty. With this guarantee of governmental support, the Dow board looked at my proposal to build two new plants in Midland, and approved it in nanoseconds. As a result, our region is now home to two renewable projects that employ 3,000 people, not to mention all the related benefits and jobs that go with the supply chain.

Looking ahead, Dow and companies like ours will have to proceed cautiously, on a case-by-case basis—because it's still not clear how serious the government is about these emerging industries. Of course, politicians talk a lot about clean energy—but their focus is usually short-term job creation. The measure of this is the lack of a policy framework designed to encourage this new industry. Without one, business leaders and investors are left to wonder, is this it? Have they done all they're going to do? Or can we expect the government's commitment to continue and grow? Where there is uncertainty, I can assure you, business investment will not follow.

Just as there would be no new plants in Midland without state and federal support, there will be no meaningful clean energy industry in America without strategic action to reduce uncertainty.

The government can accomplish this in four key ways:

1. **Make a commitment to a renewable energy standard.** Countries that are expanding their renewable energy sectors are doing so in large part because their governments have committed to lowering the country's carbon emissions by a set amount over a set time period. That sends a signal to companies that the government is serious about the issue, and that the market for renewables is real and will grow. In the United States, we have no such commitment. We need one.
2. **Put a price on carbon.** Like a renewable energy standard, a price on carbon would signal the market. It would allow companies to reduce their uncertainty by calculating exact costs over an extended period. This price should be primarily pursued through market-based mechanisms to assure the lowest cost of compliance.
3. **Make renewable energy incentives permanent.** Companies need to be able to count on government action, and need

to be able to factor it into decisions about where—and if—to build. By extending incentive programs beyond the short-term, by making them, in effect, permanent, businesses can plan for the long term.

4. **Encourage energy-related innovation.** There is a world of opportunity awaiting any country that leads in creating products that are substantially more efficient than existing ones—products that can achieve more at less cost. Private sector commitment to R&D and governmental commitment to R&D go hand in hand, as we have seen in the United States in past decades, and as we see around the world. We cannot have one without the other.

The cost of early research is extraordinarily high, often prohibitively so. The most significant strides we've taken in innovation in this country have, as a result, been largely the product of substantial federal investment in R&D. The early discoveries, the early inventions, have rarely arisen exclusively in a private-sector environment. But with a concerted national effort, these innovative ideas can find the funding—and the resources—to become viable. Government R&D investments have resulted in the early phases of countless critical innovations. Companies like Dow have then worked with the government to take on some of the risk, to further develop the products, and to ultimately commercialize them.

Other countries are now following that model. They are spending substantial sums of money in focused R&D, providing the space for innovation that has no equivalent in the private sector. Those investments attract businesses like Dow. We want to be at the center of innovation—and the center of innovation is increasingly moving offshore.

The United States should substantially increase its R&D budget, and should focus specifically on industries—like clean energy—where success in innovation has the greatest potential to result in valuable products, high-paying jobs, and sustainable economic growth.

Reduce Energy Costs. In addition to building a renewable energy sector, the United States needs policies that will help reduce the cost of energy use for all manufacturers, as a means of boosting the sector financially and attracting new investment within our borders. I see two primary ways of accomplishing this:

1. **Increase efficiency.** The price of energy is one of the highest costs most manufacturers face. We know that greater efficiency will, in all circumstances, reduce those costs. All manufacturers can—and must—take it upon themselves to maximize their energy efficiency. We at Dow take that goal very seriously, and have benefited from doing so.

   But again, government policy has a role to play. Even before that energy flows to the manufacturer, even before the manufacturing site has been built, sound policies could go a long way toward encouraging power generators and distributors to undertake cost-effective energy efficiency measures. High efficiency standards in building codes could also ensure that new manufacturing sites are constructed in ways that reduce energy intensity. That alone would provide significant savings across the board.
2. **Increase natural gas supply.** About 22 percent of the energy we use in America comes from natural gas. Demand for natural gas is expected to grow substantially over the next 20 years, while the U.S. supply of natural gas from conventional sources is expected to fall 20 percent over that same period. This will directly—and negatively— impact our energy bills.

Unlike crude oil, natural gas is very much a local product. The United States produces nearly 90 percent of the natural gas it consumes. That means we'll have to depend on our local supplies to meet this new demand. This will require a national commitment to increase our supply and a willingness to explore for natural gas in shale and other potential sources.

Infrastructure

Rebuilding our nation's infrastructure will be critical for our long-term global competitiveness. Let me be frank: it will also be enormously expensive. The American Society of Civil Engineers estimates that it would cost $2.2 trillion over five years to get our infrastructure back in shape. That bill, even as the economy improves, is not one that America can afford to pay.

But we also can't afford to let our roads and bridges and ports continue to deteriorate. The economic—and human—costs will be high indeed. Meanwhile, other countries will leapfrog America, as many are already, with state-of-the-art technologies. We cannot ignore our infrastructure problems and still expect to compete. That requires us to think strategically and act swiftly, even within fiscal constraints.

Create a National Infrastructure Bank. The money the United States already spends on infrastructure—about $400 billion a year on average—isn't spent strategically. We don't decide, as a nation, which investments are most critical.

To remedy this problem, the United States should transform the way it funds infrastructure. It should depoliticize the process, as it has done with other tricky issues, such as military base closings. America should create a national infrastructure bank, funded by

Congress and directed to provide grants, loans, and loan guarantees for strategic infrastructure projects that will make a substantial national or regional impact. President Obama proposed such a bank in the fall of 2010 and Congress should move swiftly to create it.

Europe has had a similar model for more than 50 years. The European Investment Bank has played a central role in upgrading the continent's infrastructure and in connecting the European Union. It has even turned a profit!

Practice Prevention. Most of the federal funding for infrastructure comes in the form of grants to states and local governments. To qualify, projects need to meet certain criteria. The problem is that the criteria are stacked in favor of fixing things once they're broken—rather than preventing them from breaking in the first place. As a result, Brookings and others have reported that states sit back and let infrastructure assets degrade until they need a major overhaul—one big enough to qualify for federal dollars.

In recent years, we've seen a paradigm shift in health care—the system is finally giving people incentives to stay healthy rather than to wait until they get sicker and sicker. We should apply the same basic understanding to infrastructure, and award funding to projects that meet certain maintenance requirements. That funding can be drawn from a special fund for long-term preventive maintenance.

Provide Incentives for Private Infrastructure Projects. Of course, new infrastructure—whether broadband cables, new transmission lines, or computerized ports—will be built by private companies, not government. But as is the nature of business, companies are not going to invest billions of dollars in building solely because a project serves the national interest. That is an important consideration, but businesses also have shareholders to consider and risks to weigh.

The government can play a role in aligning private interests with the public interest. As in other key areas, the United States should provide targeted tax credits and other incentives to businesses that invest in high-priority projects.

Taken together, these policy proposals represent a far-reaching agenda for advanced manufacturing. If implemented, it would spur a clean energy manufacturing sector in this country, expand advanced manufacturing in other areas, prepare our future workforce to meet the needs of those jobs, and create a modern infrastructure to facilitate efficient commerce. In short, it would take the American economy back to a point where it can sustain itself, to a point where it can create long-term job growth and even longer-term prosperity.

Still, I believe these solutions, though bold, are reasonable and practical. Nothing in these pages is beyond our national capacity. All of it can be accomplished.

It won't, of course, be easy. Transformation never is, either for companies or nations. But if we commit ourselves to the pursuit of a national economic strategy, we can once again be confident about America's ability to compete and lead. The tree will have been planted. In due time, it will grow.

Chapter 8

The Fork in the Road

Andy Grove, co-founder of Intel, published a book in the mid-1990s that discussed his experiences building a business. He discussed the kinds of challenges that companies often confront, and focused on what he described as "strategic inflection points"—moments that can make or break a business.

"A strategic inflection point," he wrote, "is a time in the life of a business when its fundamentals are about to change. That change can mean an opportunity to rise to new heights. But it can just as likely signal the beginning of the end."

That's a message that resonates for every CEO, and I'm no exception. During the past decade, Dow reached a strategic inflection point of its own. We had achieved global prominence largely through the development and manufacture of basic chemicals and plastics. As I've described, oil and gas—the wellspring of petrochemicals—were our raw materials. We rose or fell along with the rest of the energy markets. A little less than 10 years ago, it became clear to us that prices for our primary inputs had grown increasingly, unacceptably volatile.

Meanwhile, the world around us was changing: like everyone else, we were witnessing game-changing, global shifts in customer and market behaviors. These megatrends meant mega-opportunities for companies with the right capabilities in place. We knew we could be one of those companies—but were not at that point. We took a hard look at ourselves and realized we had to transform our company—not just at the margins, but at the core.

We chose to change. Over the years since, we have rebalanced our geographic position to reach the markets of greatest growth; we have shifted most of our portfolio from basics to advanced materials; and we have reoriented our corporate culture toward growth, toward the customer, and toward cutting-edge technology and innovation. As Dow's CEO, I am humbled by the hard work of our many thousands of employees—their dedication to this new vision of the company has made all our present—and future—successes possible.

The details are particular to Dow alone, but the broad outlines of our story will be familiar to anyone who runs a business, especially in these changing times. Commerce—whether local or global—is not static. Businesses face turning points periodically. So do nations—and the stakes, of course, are much higher. History judges our leaders by the decisions they make in moments like these. Indeed, we judge ourselves, as we must, by the actions we take when it counts the most.

This is, as I've argued, one of those moments for the United States.

It isn't the economic downturn that got us here. The recession only unmasked deeper, systemic problems, among them the fact that America has, for decades, neglected the things that matter most to its economic health and long-term strength. The manufacturing sector that created some of the world's most important innovations, that built the middle class, has eroded—in part due to reasons beyond our control, in part because of our own neglect. As a result, we haven't just lost millions of jobs. We haven't just lost the tools to create long-term prosperity. We've lost an important part of our national identity.

I believe we can get it back.

We can get it back if we stop accepting as inevitable the shuttering of factories and staggering job losses that have come to define manufacturing. We can reinvent the sector that was once the source of America's greatest pride—if we resolve to act, and act soon. And by "we" I mean all of us: political leaders, business leaders, and the American people.

It's become so commonplace for political candidates and elected leaders to stand on a factory floor and vow to revive American manufacturing that you'd be forgiven for thinking this was a top national priority. But our policies—or our policy vacuum—speak louder than words. For all the genuine concern about the loss of jobs, and all the solemn pledges to do something about it, the fact remains that many in Washington accept our decline. They'd never admit it—perhaps not even to themselves. But when you persist in attributing the whole problem to labor costs, then you've essentially shrugged your shoulders and said there's nothing we can do about it. Not unless America wants to get in a race to the bottom with developing countries, reducing our wages and benefits until we can compete for low-skilled jobs—something we never will and never should do.

The conventional wisdom, then, is not only wrong—as I've described—but is a dead-end.

So before we can transform the policy environment in the United States, we need to transform that way of thinking in Washington. It is said that politics is the art of the possible. If so, we need our politicians to expand their sense of what's possible. A quick look at the international landscape shows that true economic transformation is achievable—and, indeed, being achieved by other countries. It shows that manufacturing can not only survive, but thrive in the global economy. American leaders can no longer afford to accept the narrative of decline. They need to make manufacturing a priority again.

If we're going to make a national commitment to building things again, it's going to require a new approach to governing. We will not succeed without a comprehensive, national economic strategy—one that extends across policy areas, across Cabinet departments and congressional committees. One that recognizes that our economic challenges do not exist in silos. Education, energy, infrastructure, tax policy, immigration, regulation, trade—all are inextricably linked. Ad hoc, piecemeal solutions won't do.

Neither will a hands-off approach. I began this book by telling you that I am not the kind of CEO who believes that government always operates best when it operates least. And by now, you understand why. We are operating in a global economy, in which governments set many of the rules of the game. Pure free-market ideology is not obsolete. Indeed, it still drives much of our success. But it does need a reality check. Believe me, I don't want the government telling me how to run Dow. But the alternative to free enterprise is not a Soviet-style command economy. That's a false choice. And despite what some might suggest, the greatness of America—and of American companies—has never been and will never be the product of benign neglect by Washington.

Government, at its best, creates a climate in which companies can fulfill their potential. That's what American business needs. Action, not inaction. Dedicated attention, not indifference. We need partners, not adversaries, in Washington and in state capitals. We need elected leaders to assume good faith on the part of business leaders, and to proceed in a spirit of collaboration and creativity. We need political leaders who think long-term, who can look beyond the next election and do the right thing. Leaders who can inspire. Leaders who can push us to achieve great things.

I am reminded of President Kennedy's call to the American people, when he challenged the nation to go to the moon within 10 years. "We choose to go to the moon in this decade and do the other things," he famously said, "not because they are easy, but because they are hard, because that goal will serve to organize and measure the best of our energies and skills, because that challenge is one that we are willing to accept, one we are unwilling to postpone."

It was a clarion call. And in response, great progress has been made.

By engaging business in this national enterprise, President Kennedy spurred the creation of entirely new industries. Research and development originally aimed at the space program produced a broad array of innovations, from satellite navigation to water purification and medical imaging. Solar panels, ultrasound scanners, and surgical technologies, among countless other things, all grew out of that effort. President Kennedy called on America's can-do, creative spirit, and America delivered.

This is the kind of leadership we need. The kind that pushes our people to reach the furthest boundaries of what is possible.

We need a government with the determination to act.

Whether our government has the *ability* to act is another question—and an important one. The experience of the past couple years has shown that, indeed, there are times when it is still possible to get big things done. But moments like these don't come along often. Few if any of the legislative achievements of the past two years would have happened absent a severe national crisis, or without a substantial majority of the president's party in Congress. Neither of those elements are in place today. In this new reality, it is hard to imagine the President getting anything substantive accomplished until his second term, if he happens to get reelected. And even if you welcome the prospect of his defeat, you can't deny that partisanship is a pox on both houses, Democratic and Republican. It will impair the ability of Mr. Obama's successor to make progress toward vital national goals.

In an environment where it takes 60 votes to do anything, where a single senator can put an anonymous, indefinite hold on anything, it may well be impossible to act on the scale that is required. In a political system where a party that retakes power chooses to repeal, reverse, and dismantle the opposition's policies, until the pendulum swings back again, business will continue to face unmitigated uncertainty. For some, it will hasten the exodus to other countries—which, in turn, will prompt another round of finger-pointing rather than soul-searching in Washington.

It's easy to say that the system is broken, but another thing to suggest how to fix it. Clearly, the nation needs to have a serious discussion about political reform. That, however, is not my area of expertise; neither is it my focus here. I hope it will suffice to say that as the world enters a golden age of manufacturing, it's simply not acceptable for America to shrug its shoulders and say, "We'd love to be a part of it, but we can't because our politics are broken."

That's not the America I knew and looked up to from Australia. It's not the America that made me choose to come here, put down roots here. The America I love is better than that.

I'm not naïve. I don't believe transformation on this scale can be made overnight. I don't think anything's going to change just because a CEO like me puts it down on paper. Or even because the President of the United States says it from the bully pulpit.

But things have to change, and our generation is the only one that can change them in time.

Business, too, has a role to play.

I mean a role that goes beyond the basic responsibilities of business to serve their shareholders, customers, and employees, and to conduct their operations in an ethical, sustainable way. I'm talking not of duties like these, important as they are, but of the broader obligations we have to advance the national interest.

Those of us in business are not used to thinking this way, but there is a wide area of intersection between private interests and the public interest. Of course, I believe that companies like Dow serve the public interest when we simply do our jobs—when we come up with innovations that improve the quality of life, when we develop better products and deliver them to customers at better prices. But again, I think there's more we can—and must—do to apply our ingenuity, our knowhow, and our determination to the greatest challenges that confront the communities we call home. Not as an act of charity, but out of enlightened self-interest. After all, we do not do our work in a vacuum—financial, moral, or otherwise.

Aligning a single business behind great national goals can be hard, given the number of competing, pressing concerns: declining margins, next quarter's profits, the threat of a hostile takeover. Rallying an entire industry, or the private sector generally, around a common purpose is even harder, by orders of magnitude.

The public tends to think of business interests as fairly unified. Many people are familiar with the Chamber of Commerce and the Business Roundtable, of which I am a member. Groups like these represent a wide range of companies, allowing us, on certain issues, to speak with one voice. The truth, however, is more complex. As you'd expect, businesses of different types have different, often divergent, goals. Manufacturing companies like mine agree with, say, financial services companies about a lot of things, but frankly, we have very different equities at stake in the policy discussions in Washington. As monolithic as we sometimes seem, the private sector is often a house divided.

For the most part, that's just fine. Private enterprise is by its very nature competitive. We aim to win, to capture market share, to be the first and best and biggest. These things are not just a matter of pride; they're a matter of survival. But sometimes this competitive mindset puts us at cross-purposes with our natural allies. Trade associations and lobby groups are an attempt to cut against this tendency—to find, instead, the common ground that ought to exist among rival manufacturers, or farmers, or clean energy companies. That's fine, too. But the times require us to think bigger than that.

The business community needs a common purpose. More than ever before, we need to work together, across industries, to develop a common agenda for American commerce—a policy agenda that would be meaningful and beneficial to all business, regardless of sector or size. The business community needs to speak effectively with one voice on key issues, not in defining the lowest common denominator, but in pushing for policy solutions that would really contribute to creating jobs and improving sustainability across the U.S. economy. Whatever our individual missions as companies, we all share an interest in getting America back on a path toward healthy, vibrant economic growth, which will increase demand for our products, and allow our companies—and the women and men of this country—to prosper for generations.

In this book, I have tried to identify some areas for possible agreement. I've also cited many thinkers (and doers) who have terrific ideas of their own. But I'd be the first to point out that no list of agenda items, however broadly beneficial, is going to advance itself. Business needs to really persuade people—inside and outside Washington—to get behind it.

Speaking frankly, I don't see a chance of that happening until the business community stops seeing Washington automatically, in almost every case, as the enemy. Above, I urged those in government to do the same—to partner with the private sector. But that, of course, has to be a two-way street. Neither side—government or business—is solely to blame for the troubled, often toxic relationship between the two. So both sides need to take responsibility for changing the dynamic.

And businesses must proceed with an awareness that we are connected to a broader community, and are answerable to it. In an important sense, labor, local interests, nonprofits, and environmental organizations are our stakeholders, too, and business must shed its sometimes reflexive mistrust of these other groups. It isn't just the right thing to do. It's the smart thing to do. Corporate social responsibility shouldn't be an after-thought, or an exercise in public relations. It should be at the core of decisions we make.

It starts with a recognition that we're all in this together—for the very practical reason that none of us can do it alone. If political leaders can act in partnership with business, if business can act in partnership with the broader community, if all can lay down their arms and focus on a common vision, we will finally be in a position to rebuild this great country. We will finally be in a place to become globally competitive, economically strong, and confident, once again, in our future.

The most effective transformations are not top-down; they are bottom-up. Leaders need to offer direction, but they also need to listen to the people they serve. And I hope that on the issues I discuss in this book, the American people will make themselves heard.

In community meetings, radio call-in shows, and at the ballot box, the public has spoken eloquently, consistently—and at times forcefully—about its hopes, fears, and frustrations in these turbulent times. It's clear that whatever their party affiliation—if they have one—they want our elected leaders to stop playing politics and to start doing something about the gaping hole in the heart of America, the vacuum left by its once thriving manufacturing sector.

At the same time, there's ambivalence out there about manufacturing itself. As I've described, many Americans have come to see manufacturing as a relic of an earlier time, like a Model-T assembly line. They start using terms like *rust belt* to describe a vast section of the country that used to be productive. When Americans hear the word *manufacturing*, they don't think of the future. They think of the past—and of a present defined by job losses, closed factories, and a middle class in peril.

This stands in stark contrast to most other parts of the world, where manufacturing conjures thoughts of opportunity, of wealth, of growth, of promise. We cannot create a thriving culture of advanced manufacturing in the United States until the American people see things the same way—until they want those jobs and those plants and those possibilities badly enough to demand them from anyone in a position of leadership.

Those of us who are privileged to hold such a position—whether in government or business—have a role in educating the public about the value of manufacturing, the high-paying jobs it can create, the innovations it can produce. We can communicate to our children the excitement of a career as a scientist or engineer, and the American tradition of global leadership in innovation. If we do these things, more Americans will come to see manufacturing as a noble and fulfilling career, not a dead-end one; and this, in turn, will affect what politicians think and do about the issue. After all, no major problem ever gets resolved in this country unless the people tell their leaders to make it a priority.

I also ask the women and men who have retired from business—the CEOs, executives, leaders of companies big and small—to run for office. Take your depth of experience, the perspectives you've gained through your career, and apply them to public service. We need the best minds of this nation to work in government, to choose a second career in public life, so that they can fight alongside business in the quest to remake our economy.

America is at a turning point—a strategic inflection point.

Change is coming. The choice is whether to drive those changes in the right direction, or to allow forces beyond our control to determine our future. I see no reason to accept the latter as our fate. Our challenges have solutions. And America is still in a strong position to implement them. We still lead the world in scientific and technological development. Our workers are still 10 times more productive than those in China, the world's second-largest economy. We have incredible advantages, ones we can leverage to our benefit.

And, not least, America has a long and storied history of facing down challenges as great as this one, and some that were greater. This country declared its independence in the face of certain war. It expanded across the continent and overcame a Civil War. It pulled itself out of the Great Depression, defeated powerful dictators, and ushered in, for its own people and many others around the world, an era of prosperity unlike any ever known. America is, at its core, a nation of hope and optimism, of hard work and resolve.

So let the rebuilding begin. Not just of factories and communities, but of our own sense of what is achievable. It's time to get started. The future has begun.

Bibliography

Aerospace Industries Association. *A Special Report: Launching the 21st Century American Aerospace Workforce*. Arlington, Virginia: Aerospace Industries Association, December 2008. http://www.aia-aerospace.org/assets/report_workforce_1208.pdf.

American Society of Civil Engineers. *2009 Report Card for America's Infrastructure*. Reston, Virginia: American Society of Civil Engineers, March 25, 2009. www.asce.org/reportcard.

Andrew, James P., Emily Stover DeRocco, and Andrew Taylor. *The Innovation Imperative in Manufacturing: How the United States Can Restore Its Edge*. Boston: The Boston Consulting Group, March 2009. http://www.bcg.com/documents/file15445.pdf.

Applied Materials. "About Applied Materials: 40 Years of Technology Leadership." *Applied Materials*, 2010. http://fab2farm.appliedmaterials.com/about/history.htm.

———. "Applied Materials 2009 Annual Report." *Applied Materials*, 2009. http://thomson.mobular.net/thomson/7/3033/4115/.

Asia Pulse. "Green Energy Expo: Record Investments in German Renewable Energy." *Asia Pulse*, April 2010. http://w3.nexis.com/new/results/docview/docview.do?docLinkInd=true&risb=21_T10210544051&format=GNBFI&sort=RELEVANCE&startDocNo=1&resultsUrlKey=29_T10210544054&cisb=22_T10210544053&treeMax=true&treeWidth=0&csi=157371&docNo=1.

Associated Press. "Levi Strauss to Close Five plants in U.S., Canada." September 25, 2003.

Association of American Railroads. *Overview of America's Freight Railroads*. Washington, D.C.: Association of American Railroads, May 2008. http://www.aar.org/PubCommon/Documents/AboutTheIndustry/Overview.pdf.

Atkinson, Robert D. *Effective Corporate Tax Reform in the Global Innovation Economy*. Washington, D.C.: The Information Technology and Innovation Foundation, July 2009. http://www.itif.org/files/090723_CorpTax.pdf.

———. *WebMemo: Create Jobs by Expanding the R&D Tax Credit*. Washington, D.C.: The Information Technology and Innovation Foundation, January 26, 2010. http://www.itif.org/files/2010-01-26-RandD.pdf.

Avalos, George. "Sunnyvale Light Firm Bridgelux Will Move to East Bay." *San Jose Mercury News*, January 18, 2010.

Backwell, Ben. "EDPR Slashes U.S. Growth Due to Political Uncertainty." *Recharge*, February 2010. http://www.rechargenews.com/energy/wind/article207349.ece.

Beckford, Martin. "Why Was Britain the Last Major Economy to Come Out of Recession?" *Daily Telegraph*, January 27, 2010. http://www.telegraph.co.uk/finance/financetopics/recession/7079737/Why-was-Britain-the-last-major-economy-to-come-out-of-recession.html.

Bibliography

Berger, Suzanne. *How We Compete*. New York: Doubleday, 2005.

Bhanoo, Sindya. "China's Smart Grid Investments Growing." *New York Times*, February 1, 2010.

Bivens, Josh. "Shifting Blame for Manufacturing Job Loss: Effect of Rising Trade Deficit Shouldn't Be Ignored." *EPI Briefing Paper* 149. Washington, D.C.: Economic Policy Institute, April 8, 2004. http://www.epi.org/publications/entry/briefingpapers_bp149/.

———. "Truth and Consequences of Offshoring: Recent Studies Overstate the Benefits and Ignore the Costs to American Workers." *Economic Policy Institute Briefing Paper* 155. Washington, D.C.: Economic Policy Institute, August 1, 2005. http://www.epi.org/publications/entry/bp155/.

Borenstein, Seth. "Oil Slick Not Just on Gulf: Petroleum Products Permeate Daily Life from Sneakers to Milkshakes." *Minneapolis Star-Tribune*, July 11, 2010.

Bradsher, Keith. "China's Route Forward." *New York Times*, January 22, 2009.

———. "China Leading Global Race to Make Clean Energy." *New York Times*, January 30, 2010. http://www.nytimes.com/2010/01/31/business/energy-environment/31renew.html.

———. "China Drawing High-Tech Research from U.S." *New York Times*, March 17, 2010. http://www.nytimes.com/2010/03/18/business/global/18research.html.

———. "China Sees Growth Engine in a Web of Fast Trains." *New York Times,* February 12, 2010.

———. "Recovery Picks Up in China as U.S. Still Ails." *New York Times*, September 17, 2009.

Brooks, David. "Relax, We'll Be Fine." *New York Times*, April 5, 2010. http://www.nytimes.com/2010/04/06/opinion/06brooks.html.

Brown, Alan S. "Why Engineering Is Moving Offshore." *Mechanical Engineering*, web exclusive. http://memagazine.asme.org/web/Moving_Offshore.cfm.

Bureau of Labor Statistics. "Textile, Textile Product, and Apparel Manufacturing." In *Career Guide to Industries, 2010-11 Edition*. Washington,

D.C.: U.S. Department of Labor, last modified: December 17, 2009. http://stats.bls.gov/oco/cg/cgs015.htm.

———. *Employees on Nonfarm Payrolls by Major Industry Sector, 1960 to date.* Washington, D.C.: Bureau of Labor Statistics, 2010. ftp://ftp.bls.gov/pub/suppl/empsit.ceseeb1.txt.

Burrows, Dan, and Jason Kephart. "That's America: 10 Stocks Launched by Immigrants." *SmartMoney*, July 1, 2009. http://www.smartmoney.com/investing/stocks/10-companies-founded-by-immigrants/.

Burrows, Peter, Jack Kaskey, and Ian King. "How to Build an American Job." *BusinessWeek*, July 1, 2010. http://www.businessweek.com/magazine/content/10_28/b4186048377696.htm.

BusinessWeek. "Mark Pinto: Executive Profile & Biography." September 2010. http://investing.businessweek.com/research/stocks/people/person.asp?personId=8442383&ticker=AMAT:US.

Cao, Cong, Richard Suttmeier, and Denis Fred Simon. "China's 15-Year Science and Technology Plan." *Physics Today*, December 2006. http://www.levin.suny.edu/pdf/Physics%20Today-2006.pdf.

Carroll, Robert. "The Importance of Tax Deferral and a Lower Corporate Tax Rate." *Tax Foundation Special Report* 174. Washington, D.C.: Tax Foundation, February 2010. http://www.taxfoundation.org/files/sr174.pdf.

Center for American Progress. "A National Clean-Energy Smart Grid 101." Washington, D.C.: Center for American Progress, February 23, 2009. http://www.americanprogress.org/issues/2009/02/pdf/smart_grid101.pdf.

Chu, Henry. "Panels Grab Sun's Power Even Amid Clouds." *Los Angeles Times*, December 26, 2009. http://w3.nexis.com/new/results/docview/docview.do?docLinkInd=true&risb=21_T10210410322&format=GNBFI&sort=RELEVANCE&startDocNo=1&resultsUrlKey=29_T10210410327&cisb=22_T10210410326&treeMax=true&treeWidth=0&csi=144571&docNo=1.

Clinton, William J. "Remarks Announcing the National Performance Review." March 3, 1993. http://www.presidency.ucsb.edu/ws/index.php?pid=46291.

Coile, Zachary. "U.S. Far Behind in Green Tech Revolution, Senators Told." *San Francisco Chronicle*, January 8, 2009. http://articles.sfgate.com/2009-01-08/news/17195521_1_clean-energy-energy-technology-climate-change.

Computer History Museum. "1967—Turnkey Equipment Suppliers Change Industry Dynamics." *Computer History Museum* under "The Silicon Engine: A Timeline of Semiconductors in Computers." http://www.computerhistory.org/semiconductor/timeline/1967-Equipment.html.

Congressional Budget Office. *Factors Underlying the Decline in Manufacturing Employment Since 2000*. Washington, D.C.: Congressional Budget Office, December 23, 2008.

Cooper, Jr., John Milton. *Pivotal Decades*. New York: W.W. Norton & Company, 1990.

Couture, Toby D., et al. "A Policymaker's Guide to Feed-in Tariff Policy Design." National Renewable Energy Laboratory Technical Report NREL/TP-6A2-44849, July 2010. http://www.nrel.gov/docs/fy10osti/44849.pdf.

Cowan, Richard. "Deutsche Bank Spurns U.S. for Climate Investment." *Reuters*, August 11, 2010. http://www.reuters.com/article/idUSTRE67A3JK20100811.

Cowen, Tyler. "What Germany Knows About Debt." *New York Times*, July 19, 2010. http://www.nytimes.com/2010/07/18/business/18view.html?scp=1&sq=What%20Germany%20knows%20about%20debt&st=cse.

Daily Mail Reporter. "UK's Manufacturing Output Falls at Fastest Rate Since Records Began 61 Years Ago." *Mail Online*, May 12, 2009. http://www.dailymail.co.uk/news/article-1180747/UKs-manufacturing-output-falls-fastest-rate-records-began-61-years-ago.html.

Dignan, Larry. "Amazon's Kindle Price Cut: Thank Prime View's Acquisition of E-Ink." *ZDNet.com*, July 9, 2009. http://www.zdnet.com/blog/btl/amazons-kindle-price-cut-thank-prime-views-acquisition-of-e-ink/20877.

Doerr, John, and Jeff Immelt. "Falling Behind on Green Tech." *Washington Post*, August 3, 2009. http://www.washingtonpost.com/wp-dyn/content/article/2009/08/02/AR2009080201563.html.

Dolan, Kerry. "China's Threat to Clean Tech." *Forbes*, March 26, 2010. http://blogs.forbes.com/velocity/2010/03/26/china%E2%80%99s-threat-to-clean-tech/.

Dougherty, Carter. "Debate in Germany: Research or Manufacturing?" *New York Times*, August 11, 2009. http://www.nytimes.com/2009/08/12/business/global/12silicon.html.

Easterbrook, Gregg. *Sonic Boom*. New York: Random House, 2009.

Ebinger, Charles K., and John P. Banks. *Nuclear Assessment*. Washington, D.C.: The Brookings Institution, April 30, 2010. http://www.brookings.edu/articles/2010/0430_nuclear_energy_banks_ebinger.aspx.

Economist. "A special Report on America's Economy: Time to Rebalance." March 31, 2010. http://www.economist.com/node/15793036?story_id=15793036.

———. "A Special Report on Germany: Older and Wiser." March 11, 2010. http://www.economist.com/node/15641069.

———. "America's Economy: Hope at Last." March 31, 2010. http://www.economist.com/node/15816636.

———. "British Exports: Trading out of Trouble." February 18, 2010. http://www.economist.com/node/15549013?story_id=15549013.

———. "Clambering out of the Hole." January 7, 2010. http://www.economist.com/node/15214018.

———. "Export or Die." March 31, 2010. http://www.economist.com/node/15793128?story_id=15793128.

———. "Manufacturing Blues: Another One Bites the Dust." January 21, 2010. http://www.economist.com/node/15331177?story_id=15331177.

———. "Skills for the Future: The Plot So Far." March 18, 2010. http://www.economist.com/node/15719266?story_id=15719266.

———. "Taxing Companies: Choose Your Weapons." March 11, 2010. http://www.economist.com/node/15671566?story_id=15671566.

———. "The British Economy: The Pain to Come." March 25, 2010. http://www.economist.com/node/15770872?story_id=15770872.

———. “The State of Britain: How Broken Is Britain?” February 4, 2010. http://www.economist.com/node/15452811.

Edwards, Chris. “Emerald Miracle.” *National Review Online* 4, March 16, 2007. http://www.nationalreview.com/articles/220321/emerald-miracle/chris-edwards.

Engardio, Pete. “Can the Future Be Built in America?” *BusinessWeek*, September 10, 2009. http://www.businessweek.com/magazine/content/09_38/b4147046115750.htm.

Engardio, Peter. “China’s Reverse Brain Drain.” *BusinessWeek*, November 19, 2009.

Ernst & Young. *International R&D Tax Incentives*. April 2008. http://www.investinamericasfuture.org/PDFs/newRDchartRev04-04-08doc1.pdf.

Fields, Gary. “Political Uncertainty Puts Freeze on Small Business.” *Wall Street Journal*, October 28, 2009. http://online.wsj.com/article/SB125659324579108943.html.

Financial Times. “Manufacturing in North-West England.” March 2, 2010. http://www.ft.com/reports/north-west-manufacturing-2010.

Fingleton, Eamonn. “Germany’s Economic Engine.” *The American Prospect*, February 24, 2010. http://www.prospect.org/cs/articles?article=germanys_economic_engine.

First Post. “British Manufacturing: A Success Story.” http://www.thefirstpost.co.uk/47376,news-comment,news-politics,british-manufacturing-a-success-story.

Flint, Jerry. “They Can Build Them; Why Can’t We?” *Forbes*, May 28, 2009. http://www.forbes.com/2009/05/27/auto-manufacturing-detroit-business-unions.html.

Florman, Samuel C. *The Existential Pleasures of Engineering*. New York: St. Martin’s Griffin, 1976.

Fountain, Henry. “Calculating Water Use, Direct and Indirect.” *New York Times*, April 19, 2010.

Friedman, Benjamin M., ed., Jagdish Bhagwati, and Alan S. Blinder, *Offshoring of American Jobs: What Response from U.S. Economic Policy*? Boston: MIT Press, 2009.

Friedman, Thomas. "Who's Sleeping Now?" *New York Times*, January 9, 2010. http://www.nytimes.com/2010/01/10/opinion/10friedman.html.

———. "Can I Clean Your Clock?" *New York Times*, July 4, 2009. http://www.nytimes.com/2009/07/05/opinion/05friedman.html.

———. "A Word from the Wise." *New York Times*, March 2, 2010. http://www.nytimes.com/2010/03/03/opinion/03friedman.html.

———. "Elvis Has Left the Mountain." *New York Times*, January 31, 2009. http://www.nytimes.com/2009/02/01/opinion/01friedman.html.

———. "Have a Nice Day." *New York Times*, September 15, 2009. http://www.nytimes.com/2009/09/16/opinion/16friedman.html.

———. "The New Sputnik." *New York Times*, September 26, 2009. http://www.nytimes.com/2009/09/27/opinion/27friedman.html?ref=thomaslfriedman.

Galbraith, Kate. "Can Clean Energy Revive Manufacturing?" *New York Times*, May 4, 2009, Green Blog. http://green.blogs.nytimes.com/2009/05/04/can-clean-energy-revive-manufacturing/.

Gates, Bill. "Remarks to American Federation of Teachers." July 20, 2010. http://www.aft.org/pdfs/press/sp_gates071010.pdf.

———. "Speech to the National Governors Association at the Achieve Summit." February 26, 2005. http://www.nga.org/cda/files/es05gates.pdf.

Germany Trade & Invest. "Industry Report; Germany's Photovoltaic R&D Is Tops." Volume 1, 2010. http://www.gtai.com/homepage/info-service/publications/our-publications/germany-investment-magazine/vol-2010/vol-012010/industry-report-green-rd/?backlink=Back%20to%20overview.

Gomory, Ralph. "The Innovation Delusion." *Huffington Post*, March 1, 2010. http://www.huffingtonpost.com/ralph-gomory/the-innovation-delusion_b_480794.html.

Gonzales, Patrick, et al. "Mathematics Performance in the United States and Internationally." *Highlights from TIMSS 2007: Mathematics and Science Achievement of U.S. Fourth- and Eighth-Grade Students in an International*

Context. National Center for Education Statistics, Institute of Education Sciences. Washington, D.C.: U.S. Department of Education, December 2008.

———. "Science Performance in the United States and Internationally." *Highlights from TIMSS 2007: Mathematics and Science Achievement of U.S. Fourth- and Eighth-Grade Students in an International Context*. National Center for Education Statistics, Institute of Education Sciences. Washington, D.C.: U.S. Department of Education, December 2008.

Gordon, Kate, Julian L. Wong, and JT McLain. *Out of the Running? How Germany, Spain and China Are Seizing the Energy Opportunity and Why the United States Risks Getting Left Behind*. Washington, D.C.: Center for American Progress, March 2010. http://www.americanprogress.org/issues/2010/03/pdf/out_of_running.pdf.

Green, David. "Failed Public Policies Are to Blame for the Decline in Manufacturing." *Daily Telegraph*, April 21, 2009. http://www.telegraph.co.uk/comment/personal-view/5194782/Failed-public-policies-are-to-blame-for-the-decline-in-manufacturing.html.

Grove, Andrew S. *Only the Paranoid Survive*. New York: Doubleday, 1996.

Grove, Andy. "How America Can Create Jobs." *BusinessWeek*, July 1, 2010. http://www.businessweek.com/magazine/content/10_28/b4186048358596.htm.

Grueber, Martin, and Tim Studt. "Emerging Economies Drive Global R&D Growth." *R&D Magazine*, December 22, 2009. http://www.rdmag.com/Featured-Articles/2009/12/Policy-And-Industry-Global-Funding-Report-Emerging-Economies-Drive-Global-R-D-Growth/.

Hamberger, Ed. "Transit's Future? Trains and Transporters." *Washington Post*, March 23, 2009.

Hian Hou, Chua. "U.S. firm Opens S'pore Plant." *Straits Times*, March 10, 2010. http://www.straitstimes.com/BreakingNews/Singapore/Story/STIStory_500325.html.

Hira, Ron. "The Offshoring of Innovation." *EPI Briefing Paper 226*. Washington, D.C.: Economic Policy Institute, December 1, 2008. http://www.epi.org/publications/entry/bp226/.

Hounshell, David. *From the American System to Mass Production, 1800-1932.* Baltimore: Johns Hopkins University Press, 1984.

Hsuan, Amy. "Oregon's Solar Future." *The Oregonian*, February 19, 2010. http://www.oregonlive.com/business/index.ssf/2010/02/oregons_solar_future_could_hin.html.

Ikenson, Daniel J. "China's Exchange Rate Policy and Trade Imbalances." Cato Institute, April 22, 2010. http://www.cato.org/pub_display.php?pub_id=12216

Industry Week. "U.S. Department of Labor Updates Advanced Manufacturing Competency Model." May 4, 2010. http://www.industryweek.com/articles/u-s-_department_of_labor_updates_advanced_manufacturing_competency_model_21751.aspx.

International Energy Agency. *Energy Statistics for Non-OECD Countries, 2010 Edition.* Paris: IEA Publications, 2010. http://www.iea.org/publications/free_new_Desc.asp?PUBS_ID=1077.

International Trade Administration. "January 2010 Export Statistics." *U.S. Export Fact Sheet*. Office of Trade and Industry Information, International Trade Administration. Washington, D.C.: U.S. Department of Commerce, March 11, 2010. http://2001-2009.commerce.gov/s/groups/public/@doc/@os/@opa/documents/content/prod01_008993.pdf.

Issenberg, Sasha. "Same State, Different Message for Michigan's Economy." *Boston Globe*, January 14, 2008. http://www.boston.com/news/nation/articles/2008/01/14/same_state_different_message_for_michigans_economy/.

Istrate, Emilia, and Robert Puentes. *Investing for Success: Examining a Federal Capital Budget and a National Infrastructure Bank.* Washington, D.C.: Metropolitan Policy Program at the Brookings Institution, December 2009. http://www.brookings.edu/~/media/Files/rc/reports/2009/1210_infrastructure_puentes/1210_infrastructure_puentes.pdf.

Kamenetz, Anya. "The 100 Most Creative People in Business 2010 #49: Mark Pinto." *Fast Company*, 2010. http://www.fastcompany.com/100/2010/49/mark-pinto.

Kanellos, Michael. "Ex-Seagate CEO Takes Over at LED Maker Bridgelux $50 Million Raised for Factory." Green Tech Media, enterprise channel, January 13, 2010. http://www.greentechmedia.com/articles/read/ex-seagate-ceo-bill-watkins-takes-over-at-led-maker-bridgelux/.

Karkaria, Urvaksh, and Douglas Sams. "Solar Module Maker Plans Georgia Plant." *Atlanta Business Chronicle*, January 22, 2010. http://atlanta.bizjournals.com/atlanta/stories/2010/01/25/story1.html.

Karmin, Monroe W, and Jeffrey L. Sheler. "JOBS: A Million That Will Never Come Back." *U.S. News & World Report*, September 13, 1982.

Karnis, Daniel. "Navigating the R&D Tax Credit." *Journal of Accountancy* 1, March 2010. http://www.journalofaccountancy.com/Issues/2010/Mar/20092122.

Katz, Bruce, and Robert Puentes. "Rethinking the Way on Infrastructure." *The Hill*, November 19, 2009. http://thehill.com/opinion/op-ed/68799-rethinking-the-way-on-infrastructure.

Katz, Jonathan. "Manufacturing CEOs Push Cap-and-Trade System." *Industry Week*, March 18, 2008. http://www.industryweek.com/articles/manufacturing_ceos_push_cap-and-trade_system_15992.aspx.

Kennedy, John F. "Address at the University of California at Berkeley." March 23, 1962. http://www.jfklibrary.org/Historical+Resources/Archives/Reference+Desk/Speeches/JFK/003POF03Berkeley03231962.htm.

Kinzie, Susan. "U.S. Colleges Bask in Surge of Interest Among Chinese." *Washington Post*, May 1, 2009.

Klein, Ezra. "Inner-City Futurism." *American Prospect*, July 27, 2007.

Kocherlakota, Narayana. "Inside the FMOC." Speech at Northern Michigan University, August 17, 2010. http://www.minneapolisfed.org/news_events/pres/speech_display.cfm?id=4525.

Korea Herald. "Korea—U.S. Trade Pact Mired in Inertia." November 30, 2009.

Kovach, Tim. "NREL Study Finds Feed-in Tariffs Are Responsible for 75% of All Solar PV Deployments." *Environmental Leader*, August 10, 2010. http://www.environmentalleader.com/2010/08/10/nrel-study-finds-feed-in-tariffs-are-responsible-for-75-of-all-solar-pv-deployments/.

KPMG. "Corporate Tax Rates 1999–2009." *KPMG's Corporate and Indirect Tax Rate Survey 2009*. KPMG International, October 2009. http://

www.kpmg.com/Global/en/IssuesAndInsights/ArticlesPublications/Documents/KPMG-Corporate-Indirect-Tax-Rate-Survey-2009.pdf.

———. "Corporate Tax Rates Footnotes." *KPMG's Corporate and Indirect Tax Rate Survey 2009*. KPMG International, October 2009. http://www.kpmg.com/Global/en/IssuesAndInsights/ArticlesPublications/Documents/KPMG-Corporate-Indirect-Tax-Rate-Survey-2009.pdf.

———. *KPMG's Corporate and Indirect Tax Rate Survey 2009*. KPMG International, October 2009. http://www.kpmg.com/Global/en/IssuesAndInsights/ArticlesPublications/Documents/KPMG-Corporate-Indirect-Tax-Rate-Survey-2009.pdf.

Kristof, Nicholas D. "The Educated Giant." *New York Times*, May 28, 2007.

Krugman, Paul. "America Goes Dark." *New York Times,* August 8, 2010. http://www.nytimes.com/2010/08/09/opinion/09krugman.html.

LaFraniere, Sharon. "China Reins in Its Rapid Growth." *New York Times*, July 14, 2010.

Lassa, Todd. "Toyota, Chrysler Have North America's Most Efficient Plants." *Motor Trend*, June 5, 2008. http://blogs.motortrend.com/6249925/car-news/toyota-chrysler-have-north-americas-most-efficient-plants/index.html.

Lee, Jaegun. "Freight Industry Facing Challenges." *Watertown Daily Times*, June 16, 2010.

Leonard, Jeremy A. *Economic Report: A Closer Look at the U.S. Corporate Tax Burden: Economic Effects of Fundamental Reform*. Arlington, Virginia: MAPI Manufacturers Alliance, June 2010. http://www.mapi.net/Filepost/ER-701.pdf.

Leuchtenburg, William. *The Perils of Prosperity.* Chicago: University of Chicago Press, 1958.

Lewin, Tamar. "Many States Adopt National Standards for Their Schools." *New York Times*, July 21, 2010.

Line, Molly. "Recession Hits America's Solar Industry." Fox News.com, January 6, 2010, Live Shots Blog. http://liveshots.blogs.foxnews.com/2010/01/06/recession-hits-americas-solar-industry/.

Liveris, Andrew N. "Stable Natural Gas Prices Will Boost Manufacturing."

Houston Chronicle, May 1, 2010. http://www.chron.com/disp/story.mpl/editorial/outlook/6985354.html.

Lohr, Steve. "Bringing Efficiency to Infrastructure." *New York Times*, April 29, 2009.

Lyons, Daniel. "Silicon Valley Wants to Stay On the Road to Prosperity." *Washington Post*, January 27, 2009.

Lynn, Barry. *End of the Line*. New York: Doubleday, 2005.

MacGillis, Brandon, and Nicolle Grayson. *China Leads G-20 Members in Clean Energy Finance and Investment*. Philadelphia: Pew Charitable Trusts Press Release, March 24, 2010. http://www.pewtrusts.org/news_room_detail.aspx?id=57972.

Mandel, Michael. "The Real Cost of Offshoring." *BusinessWeek*, June 18, 2007. http://www.businessweek.com/magazine/content/07_25/b4039001.htm.

The Manufacturing Institute. *The Facts about Modern Manufacturing, 8th ed*. Washington, D.C.: The Manufacturing Institute, 2009. http://www.nam.org/~/media/0F91A0FBEA1847D087E719EAAB4D4AD8.ashx.

McCormack, Richard. "The Plight of American Manufacturing." *American Prospect*, December 21, 2009. http://www.prospect.org/cs/articles?article=the_plight_of_american_manufacturing.

———. ed., *Manufacturing a Better Future for America*. Washington, D.C.: Alliance for American Manufacturing, 2009.

McKinsey & Company. *The Economic Impact of the Achievement Gap in America's Schools*. McKinsey & Company, April 2009. http://www.mckinsey.com/App_Media/Images/Page_Images/Offices/SocialSector/PDF/achievement_gap_report.pdf.

McLeod, Harriet. "U.S. Falling behind in Clean-Energy Race: Chu." *Reuters*, November 30, 2009. http://www.reuters.com/article/idUSTRE5B006E20091201.

Meachem, Brian J. *Performance-Based Building Regulatory Systems: Principles and Experiences*. Inter-Jurisdictional Regulatory Collaboration Committee, February 2010. http://www.irccbuildingregulations.org/pdf/A1163909.pdf.

Michels, Spencer. "Silicon Valley Story: Applied Materials Executive Relocating to China." *The Rundown, PBS News Hour* blog, April 5, 2010. http://www.pbs.org/newshour/rundown/2010/04/silicon-valley-company-adjusts-to-changing-dynamics.html.

Miles, Moore. "7-Year-Old TREAD Act Still a Work in Progress; To Have a Place at Table, Tire Industry Groups Must Stay Vigilant," *Tire Business*, April 28, 2008.

Miller, Clare Cain. "Another Voice Warns of an Innovation Slowdown." *New York Times*, August 31, 2008.

Mitchell, Daniel J. "Corporate Taxes: America Is Falling Behind." *CATO Institute Tax and Budget Bulletin* 48. Washington, D.C.: CATO Institute, July 2007. http://www.cato.org/pubs/tbb/tbb_0707_48.pdf.

Morse, Robert. "World's Best Universities: About the Rankings." *U.S. News & World Report,* September 21, 2010. http://www.usnews.com/articles/education/worlds-best-universities/2010/09/21/about-the-worlds-best-universities-rankings-.html.

Mufson, Steven. "Asian Nations Could Outpace U.S. in Developing Clean Energy; American Markets Slump Feeds Worry." *Washington Post*, July 16, 2009. http://www.washingtonpost.com/wp-dyn/content/article/2009/07/15/AR2009071503731.html.

Muro, Mark. "Amazon's Kindle: Symbol of American Decline?" *New Republic*, February 24, 2010. http://www.tnr.com/blog/the-avenue/amazon%E2%80%99s-kindle-symbol-american-decline.

New York Times. "National Schools Standards, At Last." March 13, 2010.

———. "What the Chinese Stimulus Package Means." November 23, 2008. http://www.nytimes.com/2008/11/23/business/worldbusiness/23iht-yuan.3.18074260.html.

Norris, Floyd. "A Recovery That's Factory-Built and Gaining Speed." *New York Times*, February 5, 2010. http://www.nytimes.com/2010/02/06/business/economy/06charts.html.

NPR. "America's Crumbling Infrastructure." *NPR*, June 1, 2008. http://www.npr.org/templates/story/story.php?storyId=91041006.

Office of Senator Judd Gregg. "Wyden, Gregg Introduce 'Bipartisan Tax Fairness and Simplification Act of 2010.'" Press release, February 23,

2010. http://gregg.senate.gov/news/press/release/?id=17a33d25-df43-44cc-b336-85cf0cd523f1.

O'Grady, Sean. "Made in Britain? The Crisis in Manufacturing." *Independent*, November 4, 2008. http://www.independent.co.uk/news/business/analysis-and-features/made-in-britain-the-crisis-in-manufacturing-989933.html.

Organisation for Economic Co-operation and Development. "Chapter 5: Investing in the Knowledge Economy." *OECD Science, Technology and Industry Scoreboard 2009*. Paris: OECD Publishing, 2009. http://www.oecd-ilibrary.org/content/book/sti_scoreboard-2009-en.

———. "Executive Summary." OECD Reviews of Innovation Policy: China. Paris, OECD Publishing, September 2008. http://www.oecd.org/ dataoecd/7/45/41270116.pdf.

———. "Science and Innovation: Country Notes—Brazil." *OECD Science, Technology and Industry Outlook 2008*. Paris: OECD Publishing, October 2008. http://www.oecd.org/dataoecd/18/31/41559606.pdf.

———. "Science and Innovation: Country Notes—China." *OECD Science, Technology and Industry Outlook 2008*. Paris: OECD Publishing, October 2008. http://www.oecd.org/dataoecd/18/36/41559747.pdf.

Orszag, Peter R. "Current and Future Investment in Infrastructure." Statement before the House Committee on the Budget and the House Committee on Transportation and Infrastructure, U.S. House of Representatives. Washington, D.C.: Congressional Budget Office, May 8, 2008. http://budget.house.gov/hearings/2008/05.08orszag.pdf.

Oxford Economic Forecasting. "UK Assessment: UK recessions—anatomy, causes and risk." *Economic Outlook*, 26 2002: 5-15.

Palmer, Doug. "U.S. Could Fall behind China in Clean Energy: Locke." *Reuters*, May 22, 2010. http://www.reuters.com/article/idUSTRE64L0QD20100522.

Pence, Michael. "America Needs a Growth Strategy." *Financial Times*, July 8, 2010. http://www.ft.com/cms/s/0/675c2508-8ac1-11df-8e17-00144feab49a.html.

Peskin, Lawrence A. *Manufacturing Revolution*. Baltimore: Johns Hopkins University Press, 2003.

Petroski, Henry. *The Essential Engineer.* New York: Knopf, 2010.

Pfaff, Alexander, and Chris William Sanchirico. "Big Field, Small Potatoes: An Empirical Assessment of EPA's Self-Audit Policy." *Journal of Policy Analysis and Management* 23: 3 Summer 2004: 415-432.

Pisano, Gary, and Willy Shih. "Is the U.S. Killing Its Innovation Machine?" *Harvard Business Review* 1, November 17, 2009. http://blogs.hbr.org/hbr/restoring-american-competitiveness/.

Plafker, Ted. "Economic Stimulus a Mixed Blessing for China." *New York Times*, January 20, 2010.

Podesta, John D., et al. *The Clean-Energy Investment Agenda: A Comprehensive Approach to Building the Low-Carbon Economy.* Washington, D.C.: Center for American Progress, September 21, 2009. http://www.americanprogress.org/issues/2009/09/clean_energy_investment.html.

Politi, James. "U.S. Stimulus Shifts to Infrastructure Spending." *Financial Times*, July 11, 2010.

Pomfret, John. "China Pushing the Envelope on Science, and Sometimes Ethics." *Washington Post*, June 28, 2010. http://www.washingtonpost.com/wp-dyn/content/article/2010/06/27/AR2010062703639.html.

Pool, Sean. *How to Power the Energy Innovation Lifecycle: Better Policies Can Carry New Energy Sources to Market.* Washington, D.C.: Center for American Progress, June 2010. http://www.americanprogress.org/issues/2010/06/pdf/energy_innovation.pdf.

Pope, Justin. "Push to Double U.S. Science Grads Is Lagging." *San Francisco Chronicle,* July 15, 2008. http://articles.sfgate.com/2008-07-15/business/17170914_1_math-and-science-business-groups-ceo-and-chairman.

Porter, Michael E. "Why America Needs an Economic Strategy." *BusinessWeek,* October 30, 2008. http://www.businessweek.com/magazine/content/08_45/b4107038217112.htm.

Prest, Michael. "Making Things Worse." *Prospect Magazine,* Issue 151, October 25, 2008. http://www.prospectmagazine.co.uk/2008/10/makingthingsworse/

Reagan, Ronald. "Remarks on Signing the Tax Reform Act of 1986." October 22, 1986. http://www.americanrhetoric.com/speeches/ronaldreagantaxreformactof1986.html.

Reich, Robert. "The Future of Manufacturing: Can We Live without GM?" *Salon*, June 1, 2009. http://www.salon.com/news/ opinion/feature/2009/06/01/reich_manufacturing_gm.

Reich, Robert B. "Manufacturing Jobs Are Never Coming Back." *Forbes*, May 28, 2009. http://www.forbes.com/2009/05/28/robert-reich-manufacturing-business-economy.html.

Reuters, "$800 Million in Stimulus Will Expand Broadband." July 3, 2010. www.reuters.com

Rich, Motoko. "Factory Jobs Return, but Employers Find Skills Shortage." *New York Times*, July 1, 2010.

Riley, Bob, Haley Barbour, Ed Rendell, Phil Bredesen, and Bob McDonnell. "The Right Track for Our Roads; A Rail Corridor Would Get Traffic and the Economy Moving Again." *Washington Post*, March 26, 2010. www.wapo.com.

Rivoli, Pietra. "Sticking It to China." *Forbes*, May 28, 2009. http://www.forbes.com/2009/05/27/taratape-manufacturing-china-business-rivoli.html.

Saint, Nick. "New Survey Confirms Shoddy U.S Broadband." *Business Insider*, July 19, 2010. http://www.businessinsider.com/new-survey-confirms-that-the-us-shoddy-broadband-2010-7.

Schmit, Julie. "Demand for Upgraded Energy Efficiency at Home Is Weak." *USA Today*, January 5, 2010. http://www.usatoday.com/tech/news/2010-01-05-home-energy-efficiency-demand_N.htm.

Schneider, Howard. "Asian Nations Emerge from Recession as Stronger Economic Powers." *Washington Post*, May 14, 2010. http://www.washingtonpost.com/wp-dyn/content/article/2010/05/13/AR2010051305534.html.

Schwab, Klaus ed. "Data Tables: Section VI: Goods Market Efficiency." *The Global Competitiveness Report 2009–2010*, edited by Klaus Schwab. Geneva: World Economic Forum, 2010, 427.

http://www3.weforum.org/docs/WEF_GlobalCompetitivenessReport_2010-11.pdf.

Scott, Robert E. "China Trade and Jobs: Responding to Myths and Critics." Washington, D.C.: Economic Policy Institute, April 6, 2010. http://www.epi.org/analysis_and_opinion/entry/china_trade_and_jobs-responding_to_myths_and_critics/.

———. "Unfair China Trade Costs Local Jobs." *Economic Policy Institute Briefing Paper* 260. Washington, D.C.: Economic Policy Institute, March 23, 2010. http://www.epi.org/publications/entry/bp260.

Semiconductor Industry Association. "The Semiconductor Industry Association's Comments to the President's Economic Recovery Advisory Board's Tax Reform Subcommittee." October, 15, 2009. http://www.whitehouse.gov/assets/formsubmissions/109/df34744d431140077a53ecfb8479b7898.pdf.

Shierholz, Heidi. "Signs of Healing in the Labor Market, though Unemployment Remains in Double Digits." Washington, D.C.: Economic Policy Institute, December 4, 2009. http://www.epi.org/analysis_and_opinion/entry/signs_of_healing_in_the_labor_market_though_unemployment_remains_in_double_/.

Shih, Willy C. "The U.S. Can't Manufacture the Kindle and That's a Problem." *Harvard Business Review*, October 13, 2009. http://blogs.hbr.org/hbr/restoring-american-competitiveness/2009/10/the-us-cant-manufacture-the-ki.html.

Shin, Annys. "Toymakers Frustrated by Patchwork of Safety Rules." *Washington Post*, June 24, 2008. http://www.washingtonpost.com/wp-dyn/content/article/2008/06/23/AR2008062302163.html.

Shuman, Michael. "Does China Need a Second Stimulus?" *Time*, July 15, 2010. http://curiouscapitalist.blogs.time.com/2010/07/15/does-china-need-a-second-stimulus/.

Sissine, Fred. "Renewable Energy R&D Funding History: A Comparison with Funding for Nuclear Energy, Fossil Energy, and Energy Efficiency R&D." Washington, D.C.: Congressional Research Service RS22858, April 9, 2008.

Smick, David M. *The World Is Curved*. New York: Penguin, 2008.

Sofge, Erik. "Why Shovel-Ready Infrastructure Is Wrong." *Popular Mechanics*, February 5, 2009. http://www.popularmechanics.com/technology/engineering/infrastructure/4302578.

Splinter, Mike. "Applied Materials in Dow Jones Op-Ed: America's Next Great Industry." *Applied Materials blog*, May 25, 2010. http://blog.appliedmaterials.com/applied-materials-dow-jones-op-ed-america%E2%80%99s-next-great-industry.

Spotts, Peter N. "For China, a Reverse Brain Drain in Science." *The Christian Science Monitor*, May 1, 2009.

Stevens, Robert J. "Social Engineering." *Wall Street Journal*, April 19, 2006.

Stoessel, Amy Ann. "Recession Doesn't Ease Work Force Concerns." *Crain's Cleveland Business*, June 7, 2010.

Stoffer, Harry. "Industry Battles State-by-State MPG; Even New National Rule on Fuel Economy Could Be Tough." *Automotive News*, February 2, 2009.

Strasburg, Jenny. "Levi's to Close the Last U.S. Plants; Struggling Jeansmaker to Cut 1,980 More Jobs." *San Francisco Chronicle*, September 26, 2003. http://articles.sfgate.com/2003-09-26/business/17508000_1_levi-s-plant-closures-levi-strauss-signature-dockers.

Tankersley, Jim, and Don Lee. "China Takes Clean Lead; U.S. Falls to No. 2 in Funding for Such Alternative Sources as Wind and Solar." *Los Angeles Times*, March 25, 2010. http://articles.latimes.com/2010/mar/25/business/la-fi-energy-china25-2010mar25.

Tassey, Gregory. *The Technology Imperative*. Great Britain: MPG Books, 2007.

Tessler, Joelle. "Aerospace, Defense Sectors Brace for Brain Drain as Cold War Workers Retire." *San Francisco Chronicle*, March 10, 2008.

Theil, Stefan. "No Country Is More 'Green By Design.'" *Newsweek*, June 28, 2008. http://www.newsweek.com/2008/06/28/no-country-is-more-green-by-design.html.

Thornton, Philip. "Manufacturing Decline 'Will Double UK Trade Gap.'" *Independent*, November 1, 2001. http://www.independent.co.uk/

news/business/news/manufacturing-decline-will-double-uk-trade-gap-633387.html.

U.S. Energy Information Administration. "Emissions of Greenhouse Gases Report." December 8, 2009. http://www.eia.doe.gov/oiaf/1605/ggrpt/.

U.S. News & World Report. "World's Best Universities: Engineering and IT." September 21, 2010. www.usnews.com.

Uchitelle, Louis. "Obama's Strategy to Reverse Manufacturing's Fall." *New York Times*, July 20, 2009. http://www.nytimes.com/2009/07/21/business/economy/21manufacture.html.

United Nations Environment Programme. "Environmental Self-Auditing in the United States." http://www.unep.org/dec/onlinemanual/Enforcement/InstitutionalFrameworks/CertificationSystems/Resource/tabid/915/Default.aspx.

Vennochi, Joan. "Romney Takes Auto Industry for a Ride." *Boston Globe*, November 20, 2008. http://www.boston.com/bostonglobe/editorial_opinion/oped/articles/2008/11/20/romney_takes_auto_industry_for_a_ride/.

Versiani, Isabel, and Fernando Exman. "UPDATED-2 Brazil's Lula Unveils $878 bn Investment Plan." *Reuters*, March 29, 2010.

Washington Post. "Art of the Deal; Trade Should Be on President Obama's Agenda with South Korea." June 16, 2009.

Weber, Joseph. "Boeing to Rein in Dreamliner Outsourcing." *BusinessWeek*, January 16, 2009. http://www.businessweek.com/bwdaily/dnflash/content/jan2009/db20090116_971202.htm.

Welch, David, Dean Foust, and Coleman Cowan. "The Good News about America's Auto Industry." *BusinessWeek*, February 13, 2006. http://www.businessweek.com/magazine/content/06_07/b3971057.htm.

Welch, Joseph L. "Renewable Energy in America: Framing National Policy." Presentation at the Phase II of Renewable Energy in America National Policy Forum, Washington, D.C., December 8-9, 2009. http://www.acore.org/files/images/Welch_Joseph_L.pdf.

Wines, Michael. "China Outlines Ambitious Plan for Stimulus." *New York Times*, March 4, 2009.

Wingfield, Brian, and Daniel Indiviglio. "Manufacturers' Washington Wish List." *Forbes*, May 28, 2009. http://www.forbes.com/2009/05/27/manufacturing-obama-administration-business-washington.html.

Woelflein, Michael. "Prototype Nirvana." *Bangor Metro*, September 2006.

Wong, Edward. "China's Export Economy Begins Turning Inward." *New York Times*, June 24, 2010.

Woodhead, Michael. "How Germany Won the War for the Renewable Energy Sector." *Sunday Times London*, October 25, 2009.

World Economic Forum. "The Global Competitiveness Report 2009–2010." http://www.weforum.org/documents/GCR09/index.html.

World Education News and Reviews. "Engineering Education in India: A Story of Contrasts." New York: World Education Services, January 2007. http://www.wes.org/ewenr/07jan/feature.htm.

Wernau, Julie. "Putting Wind-Generated Power Where It's Needed." *Chicago Tribune*, March 28, 2010. http://articles.chicagotribune.com/2010-03-28/business/ct-biz-0328-wind-transmission-20100328_1_wind-power-american-wind-energy-association-wind-farms.

Xin, Hao. "China—Help Wanted: 2000 Leading Lights to Inject a Spirit of Innovation." *Science* 31 July 2009: Vol. 325. no. 5940, pp. 534–535. http://www.sciencemag.org/cgi/content/summary/325/5940/534.

Yglesias, Matt. "What's Not the Matter with American Manufacturing." *Think Progress*, December, 19, 2009. http://yglesias.thinkprogress.org/?p=38604.

Young, J.T. "YOUNG: Economy's Certain Uncertainty." *Washington Times*, January 28, 2010. http://www.washingtontimes.com/news/2010/jan/28/a-certain-uncertainty/.

Zerbe, Dean. "America's Destructive Tax Code." *Forbes*, May 28, 2009. http://www.forbes.com/2009/05/27/tax-policy-manufacturing-business-zerbe.html.

About the Author

Andrew Liveris has been the Chairman and CEO of The Dow Chemical Company since 2004. He joined the company in 1976 in Melbourne, Australia, and since then has held various production, project engineering, marketing, and business leadership positions in Australia, Hong Kong, Thailand, and the United States. His portfolio has spanned manufacturing, sales, marketing, new business development, and management. He holds a bachelor's degree in chemical engineering from the University of Queensland in Brisbane, Australia.

About The Dow Chemical Company

The Dow Chemical Company is one of the largest and most global corporations in the world. Dow is a leader in science and technology, improving quality of life and providing solutions to some of the world's biggest challenges through its many products and services.

Dow connects chemistry and innovation with the principles of sustainability to help address problems ranging from the need for clean water to renewable energy generation to the demand for increased agricultural productivity. Dow's diversified, industry-leading portfolio of specialty chemical, advanced materials, agrosciences, and plastics businesses delivers a broad mix of technology-based products and solutions to customers in approximately 160 countries. The company employs more than 52,000 people worldwide, and manufactures more than 5,000 products at 214 sites in 37 countries.

Index